AF358683

Bioactive Molecules with Healthy Features to Food and Non-food Applications

Bioactive Molecules with Healthy Features to Food and Non-food Applications

Editors

María Dolores Torres
Elena Falqué López

MDPI • Basel • Beijing • Wuhan • Barcelona • Belgrade • Manchester • Tokyo • Cluj • Tianjin

Editors
María Dolores Torres
Department of Chemical
Engineering
University of Vigo
Ourense
Spain

Elena Falqué López
Department of Analytical
Chemistry
University of Vigo
Ourense
Spain

Editorial Office
MDPI
St. Alban-Anlage 66
4052 Basel, Switzerland

This is a reprint of articles from the Special Issue published online in the open access journal *Molecules* (ISSN 1420-3049) (available at: www.mdpi.com/journal/molecules/special_issues/bio_food).

For citation purposes, cite each article independently as indicated on the article page online and as indicated below:

LastName, A.A.; LastName, B.B.; LastName, C.C. Article Title. *Journal Name* **Year**, *Volume Number*, Page Range.

ISBN 978-3-0365-1498-7 (Hbk)
ISBN 978-3-0365-1497-0 (PDF)

Contents

Preface to "Bioactive Molecules with Healthy Features to Food and Non-food Applications"

The authors of this volume provide an overview of the recent advances in the processing, characterization, structure–activity links and applications of natural bioactive molecules from a wide range of sources. The incorporation of these bioactive compounds in innovative functional matrices is also matter of interest. The chapters include discussions on Red Arils of Taxus baccata L.—a new source of valuable fatty acids and nutrients; new bioactive peptides identified from a Tilapia byproduct hydrolysate exerting effects on DPP-IV activity and intestinal hormone regulation after simulated canine gastrointestinal digestion; hydration and barrier potentials of cosmetic matrices with bee products; microwave hydrodiffusion and gravity (MHG) extraction from different raw materials with cosmetic applications; immunomodulatory effects of the Meretrix meretrix oligopeptide (QLNWD) on immune-deficient mice; marine collagen peptides promoting the cell proliferation of NIH-3T3 fibroblasts via the NF-B signaling pathway; and the impact of fermentation on phenolic compounds and the antioxidant activity of whole cereal grains: a mini review and an updated review on pharmaceutical properties of gamma-aminobutyric acid.

María Dolores Torres, Elena Falqué López
Editors

molecules

Article

Red Arils of *Taxus baccata* L.—A New Source of Valuable Fatty Acids and Nutrients

Małgorzata Tabaszewska [1], Jaroslawa Rutkowska [2,*], Łukasz Skoczylas [1], Jacek Słupski [1], Agata Antoniewska [2], Sylwester Smoleń [3], Marcin Łukasiewicz [4], Damian Baranowski [2], Iwona Duda [5] and Jörg Pietsch [6]

[1] Department of Plant Product Technology and Nutrition Hygiene, Faculty of Food Technology, University of Agriculture in Cracow, Balicka st. 122, 30-149 Cracow, Poland; malgorzata.tabaszewska@urk.edu.pl (M.T.); lukasz.skoczylas@urk.edu.pl (Ł.S.); jacek.slupski@urk.edu.pl (J.S.)

[2] Institute of Human Nutrition Sciences, Faculty of Human Nutrition, Warsaw University of Life Sciences (WULS-SGGW), Nowoursynowska st.159c, 02-776 Warsaw, Poland; agata_antoniewska@sggw.edu.pl (A.A.); damian_baranowski@sggw.edu.pl (D.B.)

[3] Department of Plant Biology and Biotechnology, Faculty of Biotechnology and Horticulture, University of Agriculture in Cracow, Al. 29 Listopada 54, 31-425 Cracow, Poland; sylwester.smolen@urk.edu.pl

[4] Department of Engineering and Machinery for Food Industry, Faculty of Food Technology, University of Agriculture in Cracow, Balicka st. 122, 30-149 Cracow, Poland; marcin.lukasiewicz@urk.edu.pl

[5] Department of Animal Product Technology, Faculty of Food Technology, University of Agriculture in Cracow, Balicka st. 122, 30-149 Cracow, Poland; iwona.duda@urk.edu.pl

[6] Institute of Legal Medicine, Medical Faculty Carl Gustav, Dresden Technical University, Fetscherstr. 74, D-01307 Dresden, Germany; joerg.pietsch@tu-dresden.de

* Correspondence: jaroslawa_rutkowska@sggw.edu.pl

Citation: Tabaszewska, M.; Rutkowska, J.; Skoczylas, Ł.; Słupski, J.; Antoniewska, A.; Smoleń, S.; Łukasiewicz, M.; Baranowski, D.; Duda, I.; Pietsch, J. Red Arils of *Taxus baccata* L.—A New Source of Valuable Fatty Acids and Nutrients. *Molecules* **2021**, *26*, 723. https://doi.org/10.3390/molecules26030723

Academic Editors: María Dolores Torres and Elena Falqué López

Received: 14 December 2020
Accepted: 27 January 2021
Published: 30 January 2021

Publisher's Note: MDPI stays neutral with regard to jurisdictional claims in published maps and institutional affiliations.

Abstract: The aim of this study, focused on the nutritional value of wild berries, was to determine the contents of macronutrients, profiles of fatty (FAs) and amino acids (AAs), and the contents of selected elements in red arils (RA) of *Taxus baccata* L., grown in diverse locations in Poland. Protein (1.79–3.80 g/100 g) and carbohydrate (18.43–19.30 g/100 g) contents of RAs were higher than in many cultivated berries. RAs proved to be a source of lipids (1.39–3.55 g/100 g). Ten out of 18 AAs detected in RAs, mostly branched-chain AAs, were essential AAs (EAAs). The EAAs/total AAs ratio approximating were found in animal foods. Lipids of RA contained seven PUFAs, including those from n-3 family (19.20–28.20 g/100 g FA). Polymethylene-interrupted FAs (PMI-FAs), pinolenic 18:3Δ5,9,12; sciadonic 20:3Δ5,11,14, and juniperonic 20:4Δ5,11,14,17, known as unique for seeds of gymnosperms, were found in RAs. RAs may represent a novel dietary source of valuable n-3 PUFAs and the unique PMI-FAs. The established composition of RAs suggests it to become a new source of functional foods, dietary supplements, and valuable ingredients. Because of the tendency to accumulate toxic metals, RAs may be regarded as a valuable indicator of environmental contamination. Thus, the levels of toxic trace elements (Al, Ni, Cd) have to be determined before collecting fruits from natural habitats.

Keywords: *Taxus baccata* L. red arils; polymethylene-interrupted fatty acids; α-linolenic acid; nutritional value; amino acids; elements

1. Introduction

Ample studies have shown fruits to be good sources of phytochemicals and to play a remarkable role in the maintenance of human health as they influence various metabolic processes [1,2]. The composition and quality of fruits derived from natural habitats depend on their genotype, but can also be modified by diverse environmental factors, such as temperature, light, water, soil quality, and altitude [3–5]. For example, temperature and light substantially determine the accumulation of soluble carbohydrates in citrus [5]. Cloudberries grown in open habitats differed significantly in their chemical composition from

Molecules **2021**, *26*, 723. https://doi.org/10.3390/molecules26030723 https://www.mdpi.com/journal/molecules

those grown in shaded sites [4]. Apart from cultivated fruits, the wild and underutilized ones also offer some nutritional value, being a rich source of carbohydrates, proteins, fibers, minerals, and vitamins [3,4,6–9].

European yew (*Taxus baccata* L.) is a non-resinous gymnosperm evergreen conifer tree or shrub up to 15 m in height, common almost all over Europe. It grows naturally at latitudes of up to 63°N in Norway and Sweden. Large populations of yew grow in Baltic countries, Ukraine, Poland, Romania, Hungary, and Carpathians and Caucasus mountains, and also in Southwest Asia and northwest Africa [10,11]. Unlike many other conifers, the common yew does not actually bear its seeds in a cone. Instead, female yews have red fleshy, berry-like structure around the seeds, known as red arils (RAs), and are open at the tip, which is equivalent to the fruit pulp of many deciduous trees. In Poland, the European yew is under species protection and is listed in the Red Book as a plant at risk of extinction [12].

The genus *Taxus* has generated considerable interest due to its content of diterpene alkaloids known as taxines [13]. Taxine B, which is detected in all parts of yew plants except RAs, is the major compound of the alkaloid fraction (approximately 30%) responsible for their toxicity. Another taxane compound, paclitaxel (taxol A), which is less polar than taxines and has cytotoxic and anticancer activities, is used in cancer therapy (lung, ovarian neoplasia, breast, metastatic carcinoma) and in the second-line treatment of AIDS-related Kaposi's sarcoma [14,15]. In addition, 10-deactetylbaccatin III, a non-alkaloidal diterpene, contains the fundamental piece of paclitaxel structure (the core taxine ring), thus inducing apoptotic cell death of cancer cells [16].

In different parts of various *Taxus* species, other active compounds such as phenolic constituents (3,5-dimethoxyphenol, myricetin, bilobetin), 50 lignans including neolignas, were identified [15]. These compounds show antibacterial, antifungal, antioxidant, and antiulcerogenic activities [17]. Strong proapoptotic activity of methanolic extract of leaves was confirmed in studies on human cell lines (colon cancer HCT, 116) [11].

It should be pointed out that in the *Taxus* species, only the seedless red fleshy part of berries, RAs, are free of toxic compounds [11,13,15]. Moreover, Siegle and Pietsch [18] revealed in the RAs of yew berries the presence of anticancer and antioxidative taxoid compounds (terpenes and phenolic compounds), with major, however trace share, of 10-deactetylbaccatin III. RAs are an enticing delicacy for many animal species, being a little expressive, slightly sweet, having a bland taste and aroma, and rich in mucous compounds. Besides, they contain a substantial amount of dietary fiber (7.7–10.6 g/100 g) [19]. However, to the best of our knowledge, published data about macro- and micro-nutrient composition of RAs is lacking.

Earlier studies on lipids of seeds of Conifer species, also of those from *Taxacea* family, showed that lipid fraction of seeds was distinguished by a substantial presence of polymethylene-interrupted fatty acids (PMI-FAs), also called Δ5-olefinic acids [20–22]. Fatty acids (FAs) from that group frequently bear the first double bond on C5, separated by five methylene units from the next double bond [23]. The chemical structure of PMI-FAs is "uncommon" as compared with polyunsaturated FAs (PUFAs) with a regular position of double bonds, e.g., linoleic and α-linolenic acids in plants [23]. However, they are typical for *Taxus* gymnosperm PMI-FAs [23]. Several pharmacological effects, e.g., modulation of immune response, suppression of hypertension, hyperlipidemia and enhancement of memory acquisition in the central nervous system, were reported for oils derived from conifer seeds containing PMI-FAs [24]. Recent studies on animal models and cell lines showed that the juniperonic acid (20:4Δ5,11,14,17) exerted anti-inflammatory effects [25]. Chen et al. [26] demonstrated that pinolenic acid (18:3Δ5,9,12) could act as a potential anti-cancer agent, reducing the risk of breast cancer by effectively antagonizing prostaglandin E_2 and cyclooxygenase expression.

Generally, fruits are not a good source of lipid fraction. However, some wild berries revealed to be a good source of lipids and beneficial FAs. For example, *Zantoxylum* fruits and wild sea buckthorn (*Hippophae rhamnoides*), contain substantial amounts of 16:1, *Rhus triparrtium* fruits having 3.8 to 6.4% lipids with substantial presence of PUFA from n-6 and n-3 families, such as *Arbutus unedo* L. berries [27–30]. Red berries of *Taxus baccata* may thus be regarded as a good source of a range of unsaturated FAs, e.g., PMI-FAs and other.

Thus, as a part of the ongoing interest in the nutritional value of wild fruits, the aim of this study was to determine the contents of macronutrients, profiles of fatty acids and amino acids, and the contents of selected elements in red arils of *Taxus baccata* L., grown in diverse locations in Poland.

2. Results and Discussion

2.1. Taxus Compounds

Five Taxus compounds were detected in RAs samples. Their contents significantly varied between collection sites (Table 1). Two of those compounds dominated: 10-deacetylbaccatin III and baccatin III, with the highest shares being found in samples from Zielona Gora sites, as compared with other three sites. This could have been due to the higher annual temperature and exposure to UV radiation at Zielona Góra site as compared with other locations (Table 2, Figure 1). The effects of light intensity and temperature on taxane concentrations in needles and twigs was reported [31]. Previous studies also revealed seasonal differentiation in taxane content, e.g., 10-deacetylbaccatin III, baccatin III and cephalomannine in seeds [14,18,32]. These non-alkaloid diterpenoid compounds are appreciated because of their proved cytotoxicity to cancer cells [13,16,33]. These results were comparable with those reported for RAs by Siegle nad Pietsch [18]. Much higher contents of 10-deacetylbaccatin III, baccatin III were detected in leaves of *Taxus baccata* L. and of other *Taxus* species (6 to 10-fold more) and in twigs of other *Taxus* species (25-fold more) [14,34].

Table 1. Taxus compounds of red arils (μg/g of dry weight, *n* = 9).

Compound	Fruit Collection Site			
	Zielona Góra	Warsaw	Koszalin	Cracow
10-Deacetylbaccatin III	19.80 ± 0.64 [c]	3.90 ± 0.13 [a]	7.40 ± 0.29 [b]	4.10 ± 0.18 [a]
Baccatin III	6.30 ± 0.13 [c]	2.00 ± 0.07 [a]	2.30 ± 0.03 [a,b]	2.40 ± 0.06 [b]
Cephalomannine	0.05 ± 0.00 [b]	0.18 ± 0.01 [c]	0.12 ± 0.01 [a]	0.12 ± 0.00 [a]
Taxinine M	0.13 ± 0.01 [d]	0.05 ± 0.00 [c]	0.03 ± 0.00 [b]	0.02 ± 0.00 [a]
Taxol A	0.02 ± 0.00 [b]	0.10 ± 0.00 [a]	0.05 ± 0.00 [c]	0.05 ± 0.00 [c]

[a, b, c, d]—values baring the same superscripts in rows do not differ significantly ($p < 0.05$) from each other.

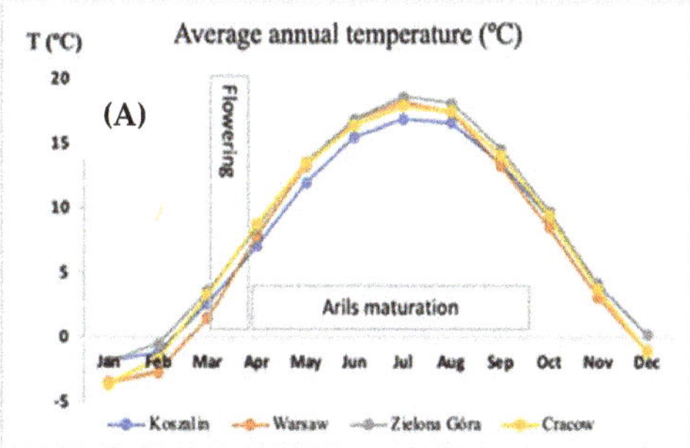

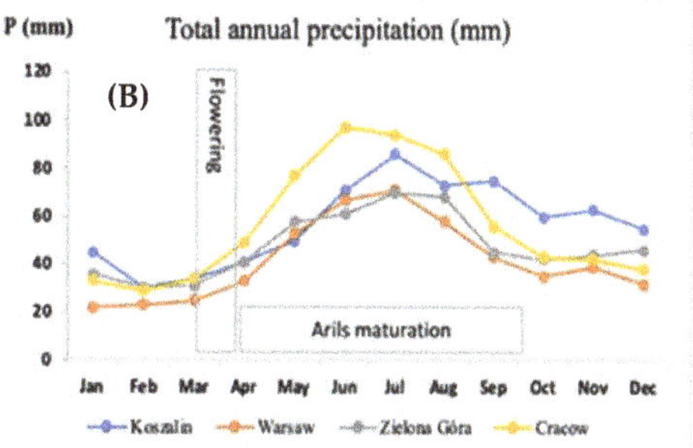

Figure 1. Annual changes in selected climate conditions in diverse locations in Poland: (**A**) average temperature (°C), (**B**) total precipitation (mm).

Table 2. Characteristics of growth locations of red arils.

Growth Site Characteristics				
Exposition	**West**	**Central**	**North**	**South**
Location of natural habitats	Zielona Góra	Warsaw	Koszalin	Cracow
Latitude/longitude	51°56′N 15°30′E	52°13′N 21°01′E	54°11′N 16°11′E	50°05′N 19°58′E
Altitude above sea level	140 m	90 m	40 m	240 m
Soil Parameters				
Type of soil	brown earths soil	podsolic soil	acidic brown soil	alluvial soil
Soil pH	>5.6	5.26–5.5	<4.5	5–5.25
Mineral nitrogen (mg/kg)	75.1–113.2	<10	20.1–35	20.1–35
Phosphorus availability (mg P_2O_5/100 g)	15–20	10–15	>25	10–15
Potassium availability (mg K_2O/100 g)	15–20	<15	<15	15–20
Magnesium availability (mg Mg/100 g)	6–7	<5	5–6	>10
Weather Conditions				
Average annual temperature (°C)	8.8	7.7	7.9	8.2
Average temperature (°C) from flowering to sampling *	13	12.1	11.7	12.6
Total annual precipitation (mm)	572	501	683	678
Average precipitation (mm) from flowering to sampling *	416	385	490	536
Cloudiness (okta)	5.6	5.7	5.9	5.3
Exposure to solar radiation (h)	1700–1800	1700–1800	1700–1800	1600–1700
Intensity of light (W/m^2)	120–140	120–140	110–130	110–130

°,*—from flowering (March–April) until the end of red arils maturation (September–October).

Other three compounds, especially taxol A (paclitaxel) were detected only in trace amounts in RAs samples (0.02–1 µ/g) in contrast with other parts of *Taxus baccata* plant, e.g., leaves, which contained 12.8 to 33.7 µ/g of paclitaxel, depending on the month of collection [14]. In leaves and twigs of other *Taxus* species, the range of amounts of paclitaxel amounted to 23.8–150 µ/g. It is worth mentioning that in the studied RAs samples, no traces of taxine alkaloids responsible for toxicity (taxines A and B) were detected. We thus consider recommending analyzing the nutritional composition of RAs as a fruit of potential value in human nutrition.

2.2. Proximate Composition

The results (Table 3) revealed a significant ($p < 0.05$) variation in the nutrient contents of RAs collected from different sites. As pointed out by Hegazy et al. [9], the moisture content affects many physical properties of fruits, such as viscosity, weight, and density, and is a helpful indicator during fruit harvesting, storage, and processing. RA samples had a higher moisture content (75.8%, on average) than wild berries from the Mediterranean region: strawberry-tree berries, blackthorn, and rose (48.7–60.9%), but lower than cultivated fruits: cherries and red raspberries (86.4–92.7%) [6,35]. As shown in Table 3, significant differences ($p < 0.05$) were noted in the fruit moisture from different locations, most likely due to different environmental conditions, such as water availability, sunlight, and wind exposition [36].

The protein content of RA (1.79–3.80%) was higher than in cultivated fruits (0.48–1%; cherry, blueberry, strawberry, red raspberry) [35]. In addition, RAs exceeded some species of wild fruits, e.g., mulberries, in protein content, but was about twice lower than in *Rhus tripartitum* fruits derived from two different locations in Tunisia [2,27].

Site differentiation in protein content of RAs was confirmed in previous papers on berry fruits and strawberry-tree fruits; it was suggested that protein content of fruits can vary with soil and climatic conditions [27,36].

Table 3. Proximate composition of red arils (means $\pm$ SE, $n = 9$).

Component	Fruit Collection Site			
	Zielona Góra	**Warsaw**	**Koszalin**	**Cracow**
Macronutrients (g/100 g of fresh weight)				
Proteins	1.79 $\pm$ 0.02 [a]	3.80 $\pm$ 0.20 [c]	3.03 $\pm$ 0.10 [b]	1.95 $\pm$ 0.17 [a]
Lipids	1.39 $\pm$ 0.12 [b]	0.79 $\pm$ 0.04 [a]	3.55 $\pm$ 1.24 [c]	3.54 $\pm$ 1.11 [c]
Carbohydrates	18.49 $\pm$ 0.13 [a]	19.06 $\pm$ 0.16 [a]	19.30 $\pm$ 0.22 [a]	18.43 $\pm$ 0.10 [a]
Glucose	2.73 $\pm$ 0.09 [b]	3.03 $\pm$ 0.12 [c]	2.23 $\pm$ 0.12 [a]	2.22 $\pm$ 0.02 [a]
Fructose	5.78 $\pm$ 0.19 [a,b]	6.43 $\pm$ 0.27 [b]	5.55 $\pm$ 0.33 [b]	5.36 $\pm$ 0.04 [a]
Sucrose	2.28 $\pm$ 0.07 [c]	1.65 $\pm$ 0.05 [b]	0.92 $\pm$ 0.03 [a]	2.65 $\pm$ 0.09 [d]
Moisture (%) *	77.90 $\pm$ 0.09 [c]	75.89 $\pm$ 0.0.07 [b]	73.63 $\pm$ 0.08 [a]	75.71 $\pm$ 0.11 [b]
Dry matter (%) *	22.10 $\pm$ 0.06 [a]	24.11 $\pm$ 0.17 [b]	26.37 $\pm$ 0.05 [c]	24.29 $\pm$ 0.03 [b]
Ash (%) *	0.43 $\pm$ 0.01 [b]	0.46 $\pm$ 0.01 [b,c]	0.49 $\pm$ 0.01 [c]	0.37 $\pm$ 0.02 [a]
Energy value (kcal)	93.63 $\pm$ 0.15 [a]	98.55 $\pm$ 0.21 [b]	121.27 $\pm$ 0.24 [d]	113.38 $\pm$ 0.12 [c]

[a, b c, d]—values baring the same superscripts in rows do not differ significantly ($p < 0.05$) from each other. *—from flowering (March–April) until the end of red arils maturation (September–October).

Carbohydrates were the main macronutrient in RAs, accounting for 18.43 to 19.30 g/ 100 g, i.e., much more than in cultivated berries, such as strawberries, red raspberries, and blueberries (6.30–11.54 g/100 g), and in mulberries [2,35]. In turn, a higher content of carbohydrates than in RAs was assayed in wild berries [6]. The chromatographic analysis revealed the predominant share of fructose (5.36–6.43 g/100 g) followed by glucose and sucrose, irrespective of collection site (Table 4). Similar quantitative shares of sugars were confirmed in different wild berries (strawberry-tree, blackthorn, rose fruits) [6]. The sucrose content of RAs was significantly ($p < 0.05$) differentiated between all locations. The lowest content of sucrose (0.92 g/100 g), together with highest amount of fiber (10.6 g/100 g), was found in RAs collected at the Koszalin site [19]. Samples from the other three sites contained higher amounts of sucrose (1.65–2.65 g/100 g), most likely because of differences in temperature during ripening (Table 2). Our results contrasted with the data of Zheng et al. [5] about the contents of soluble sugars in *Lycium barbarum* berries; they found that genetic factors and the degree of maturity had a larger effect on sugar contents than environmental factors.

Table 4. Amino acids composition (mg/100 g of fresh weight) of red arils (means $\pm$ SE, $n = 9$).

Amino Acids	Fruit Collection Site			
	Zielona Góra	**Warsaw**	**Koszalin**	**Cracow**
Essential amino acids				
Histidine	80.0 $\pm$ 0.0 [c]	90.0 $\pm$ 0.0 [d]	43.3 $\pm$ 3.3 [a]	50.0 $\pm$ 0.0 [b]
Threonine	80.0 $\pm$ 0.0 [b]	110.0 $\pm$ 0.0 [c]	80.0 $\pm$ 0.0 [b]	73.3 $\pm$ 3.3 [a]
Tyrosine	40.0 $\pm$ 0.0 [b]	53.3 $\pm$ 3.3 [c]	30.0 $\pm$ 0.0 [a]	30.0 $\pm$ 0.0 [a]
Valine	130.0 $\pm$ 0.0 [c]	166.7 $\pm$ 3.3 [d]	106.7 $\pm$ 3.3 [b]	90.0 $\pm$ 0.0 [a]
Methionine and cysteine	100.0 $\pm$ 0.0 [a]	100.0 $\pm$ 0.0 [a]	100.0 $\pm$ 0.0 [a]	100.0 $\pm$ 0.0 [a]
Lysine	110.0 $\pm$ 0.0 [a]	170.0 $\pm$ 0.0 [c]	143.3 $\pm$ 3.3 [b]	110.0 $\pm$ 0.0 [a]
Isoleucine	100.0 $\pm$ 0.0 [c]	120.0 $\pm$ 0.0 [d]	73.3 $\pm$ 3.3 [b]	63.3 $\pm$ 3.3 [a]
Leucine	183.3 $\pm$ 3.3 [c]	220.0 $\pm$ 0.0 [d]	133.3 $\pm$ 3.3 [b]	123.3 $\pm$ 3.3 [a]
Phenylalanine	173.3 $\pm$ 3.3 [c]	176.7 $\pm$ 3.3 [c]	70.0 $\pm$ 0.0 [a]	80.00 $\pm$ 0.0 [b]
Total	996.6 $\pm$ 0.7 [c]	1206.7 $\pm$ 1.0 [d]	779.9 $\pm$ 1.7 [b]	719.9 $\pm$ 1.0 [a]
Non-essential amino acids				
Aspartic acid	90.0 $\pm$ 0.0 [a]	143.0 $\pm$ 3.3 [c]	147.0 $\pm$ 3.3 [c]	113.0 $\pm$ 3.3 [b]
Serine	170.0 $\pm$ 0.0 [b]	236.7 $\pm$ 3.3 [c]	153.3 $\pm$ 3.3 [a]	153.3 $\pm$ 3.3 [a]
Glutamic acid	200.0 $\pm$ 0.0 [a]	510.0 $\pm$ 5.7 [d]	316.7 $\pm$ 6.7 [c]	283.3 $\pm$ 3.3 [b]
Glycine	126.7 $\pm$ 3.3 [c]	146.7 $\pm$ 3.3 [d]	80.0 $\pm$ 0.0 [a]	90.0 $\pm$ 0.0 [b]
Arginine	70.0 $\pm$ 0.0 [a]	100.0 $\pm$ 0.0 [c]	90.0 $\pm$ 0.0 [b]	90.0 $\pm$ 0.0 [b]
Alanine	110.0 $\pm$ 0.0 [a]	226.7 $\pm$ 3.3 [c]	236.7 $\pm$ 6.7 [c]	203.3 $\pm$ 3.3 [b]
Proline	150.0 $\pm$ 0.0 [a]	220.0 $\pm$ 0.0 [b]	210.0 $\pm$ 5.8 [b]	153.3 $\pm$ 3.3 [a]
Total	916.7 $\pm$ 0.5 [a]	1583.1 $\pm$ 2.7 [d]	1233.7 $\pm$ 3.7 [c]	1086.2 $\pm$ 2.4 [b]

[a, b c, d]—values baring the same superscripts in rows do not differ significantly ($p < 0.05$) from each other.

Generally, RAs contained a substantial amount of lipids, reaching 1.39 to 3.55 g/100 g (except those from Warsaw site). Our results opposed the view that fruits were generally a poor lipid source, such as *Rosa rugosa* pericarp (0.67–0.88%), mulberry species (0.14–0.40%), blackberries, red raspberries, and strawberries (0.25–0.42%) [2,35,37]. However, berries of some species: Goji berries, wild fruits from Saudi Arabia, are rich in lipids (2.23–5.5%) [9,38]. Total carbohydrates, simple sugars, and proteins are the vital nutrients in many fruits, as they are the main source of energy [9]. Our results also proved RAs to be an important source of lipids.

The marked differentiation in lipid contents in RAs samples derived from different sites is in agreement with previously reported data concerning other berries, namely wild *Arbutus unedo* (0.72–1.66%), mulberries (0.14–0.40%) [2,30]. That discrepancy may be due to different environmental conditions [30].

On the basis of the proximate analysis, a portion of 100 g of RAs provides, on average, 106 kcal. It is about 3.5-fold lower than the energy value of other wild berries (strawberry-tree, blackthorn, rose), most likely because of its low sugar content [6]. Hence, RAs can be recommended as a low-calorie snack.

The ash content of RAs, reaching 0.44 g/100 g (Table 3), was about twice higher than that in cultivated fruits (blackberries, red raspberries, and strawberries) and similar to that of *Prunus avium* L. [35,39]. The contents of both dry matter and ash in plants are usually affected by the climate and soil conditions [3].

2.3. Amino Acid Profile

Eighteen AAs were detected in RAs samples. Ten of them were essential AAs (EAAs), and seven non-essential AAs (non-EAAs). The contents of AAs varied significantly ($p < 0.05$) depending on the growth site (Table 4). For example, the differences in total EAAs were remarkable, ranging from 720 mg/100 g to 1207 mg/100 g (Cracow and Warsaw sites, respectively). A similar range of total EAAs (TAAs) was found in berries of Rosa *roxburghii* and Rosa *sterilis* and *Rhoodomyrtus tomentosa* (sim fruits) [8,40]. Environmental variation in AAs was also found in goji berries in China. The authors stated that alkaline soil and large day/night temperature difference were optimal for wolfberry fruit production [7].

The EAAs/TAAs ratio in RA amounted to 43%, on average. According to FAO and WHO, foods with EAAs/TAAs ratio above 40% are an ideal protein source. Generally, foods of animal origin, such as eggs, milk, and fish, have EAAs/TAAs ratios of 41 to 46% and are sources of high-quality protein [2]. The EAAs/TAAs ratio of RA approximates that found in animal foods, suggesting that the RAs could be also used as a source of high-quality protein in human diet.

Leucine was the major component of EAAs, followed by lysine and valine. Predominating shares of lysine and leucine were also found in wild fruits collected from Saudi Arabia [9]. The contents of leucine in RA (123–220 mg/100 g) were 2.5 to 3-fold higher than in Rosa *roxburghii* and Rosa *sterilis* berries [8]. The content of lysine, which is especially important for growing organisms, was similar to that found in rose hips [8]. The content of branched-chain AAs (BCAAs: leucine, isoleucine, and valine) in food is especially important, as these are closely related to human health. Apart from being the building blocks of proteins, they control the protein and energy metabolism, and serve as amino-group donors to synthesize glutamate in brain [41]. The share of BCCAs in total AAs of RAs was impressive, reaching 18.4% on average, and was similar to that found in animal proteins (about 20%) [41]. Hence, we suppose that RA can serve as a BCCAs-rich new ingredient in diet.

Glutamic acid was the major component of the non-EAAs, with 1.5 to 2.5-fold higher content in RAs from Warsaw site (510 mg/100 g) compared with samples from the other three sites (200–317 mg/100 g). The prevailing share of glutamic acid in AAs of wild fruits was also confirmed by Guo et al. [7]. The major AAs from the non-EAAs group in RA included serine (153–237 mg/100 g) and proline (150–220 g/100 g).

2.4. Fatty Acid Composition

The application of high-resolution GC enabled detecting 25 fatty acids (FAs) in lipid fraction extracted from RAs. The contents of many of them differed significantly ($p < 0.05$) depending on the fruit collection site (Table 5). Saturated FAs (SFAs) were represented mainly by palmitic (20.43–24.37 g/100 g FA) and myristic (6.76–10.76 g/100 g FA) acids. The total content of other six SFAs was low and did not exceed 3 g/100 g FA. A similar content of C16:0 was found in *Rosa rugosa*, but a much higher one in sea buckthorn pericarps [42]. The presence of palmitic and myristic acids in *Taxaceae* species (seeds) was confirmed in a previous study; however, their contents were much lower than in RA from our study [22].

Table 5. Fatty acid composition (g/100 g of FA) of lipids of red arils (means ± SE, $n = 9$).

Fatty Acid	Fruit Collection Site			
	Zielona Góra	Warsaw	Koszalin	Cracow
SFAs	33.33 ± 0.14 [a]	38.39 ± 0.15 [b]	33.12 ± 0.31 [a]	38.51 ± 0.04 [b]
10:0	0.05 ± 0.00 [a]	0.12 ± 0.01 [b]	0.07 ± 0.00 [a]	0.13 ± 0.00 [c]
12:0	0.50 ± 0.01 [b]	1.13 ± 0.01 [c]	0.30 ± 0.02 [a]	0.51 ± 0.02 [b]
14:0	9.84 ± 0.06 [b]	10.76 ± 0.05 [d]	6.76 ± 0.04 [a]	10.39 ± 0.07 [c]
16:0	20.43 ± 0.10 [a]	22.66 ± 0.29 [b]	22.37 ± 0.31 [b]	24.37 ± 0.10 [c]
17:0	0.14 ± 0.01 [c]	0.12 ± 0.01 [b]	0.13 ± 0.00 [b,c]	0.08 ± 0.00 [a]
18:0	1.88 ± 0.02 [a]	2.62 ± 0.07 [c]	3.37 ± 0.03 [d]	2.19 ± 0.03 [b]
22:0	0.16 ± 0.00 [a]	0.18 ± 0.00 [d]	0.11 ± 0.01 [c]	0.24 ± 0.01 [b]
24:0	0.32 ± 0.02 [b]	0.81 ± 0.03 [d]	0.00 ± 0.00 [a]	0.59 ± 0.01 [c]
MUFAs	10.71 ± 0.07 [c]	8.71 ± 0.08 [b]	13.04 ± 0.04 [d]	7.69 ± 0.06 [a]
14:1	0.13 ± 0.00 [a]	0.16 ± 0.02 [c]	0.15 ± 0.00 [b]	0.26 ± 0.01 [d]
16:1Δ7	0.12 ± 0.01 [a]	0.27 ± 0.00 [b]	0.12 ± 0.00 [a]	0.35 ± 0.01 [c]
16:1Δ9	0.20 ± 0.01 [b]	0.29 ± 0.00 [d]	0.26 ± 0.00 [c]	0.16 ± 0.00 [a]
17:1	0.11 ± 0.01 [a]	0.85 ± 0.03 [c]	0.14 ± 0.01 [a]	0.28 ± 0.02 [b]
18:1Δ9	9.52 ± 0.06 [b]	6.65 ± 0.08 [a]	12.35 ± 0.06 [c]	6.59 ± 0.10 [a]
18:1Δ11	0.39 ± 0.02 [c]	0.37 ± 0.03 [c]	0.00 ± 0.00 [a]	0.07 ± 0.00 [b]
20:1Δ9	0.24 ± 0.01 [d]	0.11 ± 0.00 [c]	0.04 ± 0.00 [b]	0.00 ± 0.00 [a]
***n*-3-PUFAs**	19.19 ± 0.17 [a]	25.85 ± 0.11 [c]	24.18 ± 0.20 [b]	28.19 ± 0.17 [d]
18:3Δ9,12,15	18.53 ± 0.18 [a]	25.18 ± 0.14 [c]	23.43 ± 0.23 [b]	26.50 ± 0.19 [d]
20:3Δ11,14,17	0.67 ± 0.03 [a]	0.67 ± 0.01 [a]	0.76 ± 0.02 [b]	1.69 ± 0.02 [c]
***n*-6-PUFAs**	33.81 ± 0.16 [c]	24.24 ± 0.20 [b]	25.19 ± 0.16 [b]	21.00 ± 0.13 [a]
18:2Δ9,12	30.92 ± 0.12 [c]	20.99 ± 0.22 [b]	21.33 ± 0.11 [b]	19.40 ± 0.11 [a]
18:3Δ6,9,12	0.63 ± 0.03 [b]	0.71 ± 0.01 [c]	0.80 ± 0.02 [d]	0.50 ± 0.02 [a]
20:2Δ11,14	0.16 ± 0.01 [a]	0.14 ± 0.00 [a]	0.16 ± 0.00 [a]	0.62 ± 0.02 [b]
22:2Δ13,16	1.62 ± 0.03 [b]	1.96 ± 0.01 [c]	2.33 ± 0.04 [d]	0.07 ± 0.00 [a]
20:4Δ8,11,14,17	0.48 ± 0.02 [b]	0.44 ± 0.01 [a,b]	0.57 ± 0.02 [c]	0.41 ± 0.01 [a]
PMI-FAs	1.41 ± 0.01 [b]	1.32 ± 0.01 [b]	1.08 ± 0.00 [a]	0.98 ± 0.01 [a]
18:3Δ5, 9,12	0.09 ± 0.01 [a]	0.14 ± 0.00 [b]	0.08 ± 0.00 [a]	0.23 ± 0.01 [c]
20:3Δ5,11,14	1.25 ± 0.03 [d]	0.94 ± 0.02 [c]	0.82 ± 0.02 [b]	0.62 ± 0.02 [a]
20:4Δ5,11,14,17	0.06 ± 0.00 [b]	0.24 ± 0.00 [b]	0.17 ± 0.00 [a]	0.12 ± 0.00 [c]
PUFAs	54.41 ± 0.23 [c]	51.40 ± 0.12 [b]	50.45 ± 0.16 [a]	50.16 ± 0.08 [a]
NI	1.55 ± 0.32 [a]	1.50 ± 0.21 [a]	3.38 ± 0.45 [b]	3.63 ± 0.15 [b]

[a, b c, d]—values bearing the same superscripts in rows do not differ significantly ($p < 0.05$) from each other. Abbreviations: PMI-FAs—polymethylene-interrupted FAs; 18:3Δ5,9,12—pinolenic acid; 20:3Δ5,11,14—sciadonic acid; 20:4Δ5,11,14,17—juniperonic acid; NI—not identified.

The total content of MUFAs RAs lipids varied from 7.69 to 13.04 g/100 g FA. The same trend was reported in the content of the major MUFA—oleic acid. In contrast to the lipid composition of other berry fruits, MUFAs were the least significant FAs in the RAs [38,42,43]. Lipid fraction of RAs from Koszalin site had about twice higher content of oleic acid compared, with samples from Cracow and Warsaw sites (Table 2). It is in accordance with the study of Issaoui et al. [44], who found that Tunisian olive oils from the north locations showed greater content of oleic acid comparing with samples from the south.

The lipid fraction of RAs was extremely rich in PUFAs (10 compounds). Among them, five compounds belonged to the n-6 family, and two to the n-3 family. The major PUFA was linoleic acid (30.92 g/100 g FA); it was the most abundant in the lipid fraction of Zielona Góra-site samples. The α-linolenic was the second important PUFA, especially in samples from the other three locations (23.43–26.50 g/100 g FA in average; Table 5). It should be noted that similar to *Rosa rugosa* fruits, the RAs may be perceived as a valuable source of α-linolenic acid belonging to the n-3 family [37]. Lipids of RAs also contained a long-chain PUFA (LC-PUFA) trienoic acid (20:3Δ11,14,17) from the n-3 family (Table 5), the necessary precursor of juniperonic acid [23]. That compound is rather unusual for vegetable oils.

Among other PUFAs, three PMI-FAs (18:3Δ5,9,12; C20:3Δ5,11,14 and 20:4Δ5,11,14,17), unique for the gymnosperm plants, were identified [20–23]. Even though pinolenic acid does not belong to essential FAs, it forms biologically active metabolites in the presence of cyclooxygenase or lipoxygenase, and these metabolites can partially relieve some of the symptoms of essential FAs deficiency [45]. Focusing on PMI-FAs, sciadonic dominated quantitatively, its content in RAs lipids from Zielona Góra site being higher compared with other sites. The highest content of pinolenic acid was found in lipids of RA derived from Cracow site (0.23 g/100g FA). Those results are in accordance with previously reported in lipids of *Taxus baccata* seeds [20,22]. However, our study did not confirm the presence of taxoleic acid (18:2Δ5,9), a typical FA for lipids of *Taxus baccata* seeds (Table S1) [20–22]. Substantial differences between the FAs profile of pericarp and seeds of fruits confirmed a previous study on *Garcinia* fruits [46]

There are several factors that can affect FAs composition including plant origin, environmental conditions and temperature, throughout the time between flowering and ripening [47]. For example, low temperature promotes the synthesis of PUFAs in plants, especially the LC-PUFA ones [29]. This was noted in lipids of RAs from Koszalin site, having the highest share of four LC-PUFA among studied samples (Table 5).

Contrary to expectations, lipids of RAs samples derived from Zielona Góra site, which had better climatic conditions (temperature) than other locations, contained a significantly higher amount of total PUFA (54.4 g/100 g FA), as compared with samples of the other three sites (50.7 g/100 g FA on average; Table 5), most likely due to better parameters of the brown soil there (Table 2). An impressive amount of α-linolenic acid in RAs from the Cracow sample could be attributed to favorable soil conditions and high precipitation, appropriate for the demanding *Taxus baccata* trees [48].

For nutritional reasons, it is essential to search for sustainable vegetable sources of PUFAs, especially for α-linolenic acid from the n-3 family. The primary biological role of α-linolenic may consist of it being a substrate for long chain PUFAs in EPA and DHA synthesis [49,50]. Baker et al. [49] pointed out in their review that epidemiological studies in Europe, USA, and Japan indicated a decreased risk of CVD and inflammation with increasing consumption of long-chained n-3 PUFAs. The lack of α-linolenic provision in the diet decreases the availability of DHA for incorporation into neural and retinal membranes and may explain the impact of α-linolenic deficiency on vision [50]. From the consumer's point of view and for nutritional reasons, the contents of PUFA were computed per 100 g of RAs (Table 6). The differences in PUFAs contents were related mainly to lipid content (Table 3), thus the portion of 100 g RAs from Koszalin and Cracow sites differed by a 2.5 to 4-fold higher content of PUFA compared with samples from the other two sites. In addition, RA from Koszalin and Cracow sites had high amounts of unique PMIs, especially sciadonic (on average, 25.5 mg/100 g RAs; Table 6).

Table 6. The content of important fatty acids of red arils (mg/100 g of fresh weight; means ± SE, $n = 9$).

Fatty Acid	Fruit Collection Site			
	Zielona Góra	Warsaw	Koszalin	Cracow
n-3-PUFAs	267.79 ± 1.73 [b]	204.19 ± 0.85 [a]	858.51 ± 6.77 [c]	997.81 ± 5.12 [d]
18:3Δ9,12,15	257.52 ± 2.05 [b]	198.90 ± 0.91 [a]	831.65 ± 6.62 [c]	937.98 ± 5.43 [d]
20:3Δ11,14,17	9.27 ± 0.33 [b]	5.29 ± 0.06 [a]	26.86 ± 0.54 [c]	59.83 ± 0.60 [d]
n-6-PUFAs	469.96 ± 1.35 [b]	191.47 ± 1.46 [a]	894.36 ± 3.86 [d]	743.40 ± 3.74 [c]
18:2Δ9,12	429.74 ± 1.31 [b]	165.82 ± 1.44 [a]	757.33 ± 3.28 [d]	686.64 ± 3.08 [c]
18:3Δ6,9,12	8.76 ± 0.36 [b]	5.61 ± 0.04 [a]	28.40 ± 0.60 [d]	17.70 ± 0.44 [c]
20:2Δ11,14	2.27 ± 0.08 [b]	1.08 ± 0.02 [a]	5.80 ± 0.10 [c]	21.83 ± 0.59 [d]
22:2Δ13,16	22.56 ± 0.36 [c]	15.51 ± 0.04 [b]	82.72 ± 1.10 [d]	2.60 ± 0.10 [a]
***n*-6/*n*-3**	**1:1.7 [d]**	**1:1 [b]**	**1:1 [b]**	**1:0.8 [a]**
PMI-Fas *	19.55 ± 0.46 [b]	10.43 ± 0.10 [a]	38.22 ± 0.59 [d]	34.57 ± 0.42 [c]
18:3Δ5,9,12	1.25 ± 0.07 [a]	1.13 ± 0.02 [a]	2.96 ± 0.10 [b]	8.14 ± 0.33 [c]
20:3Δ5,11,14	17.42 ± 0.36 [b]	7.43 ± 0.13 [a]	29.23 ± 0.42 [d]	22.07 ± 0.42 [c]
20:4Δ5,11,14,17	0.88 ± 0.04 [a]	1.87 ± 0.06 [b]	6.04 ± 0.17 [d]	4.37 ± 0.10 [c]
Total PUFAs	756.30 ± 3.26 [b]	406.09 ± 0.94 [a]	1791.09 ± 5.78 [c]	1775.78 ± 2.84 [c]

[a, b, c, d]—values bearing the same superscripts in rows do not differ significantly ($p < 0.05$) from each other. Abbreviations: PMI-FAs—polymethylene-interrupted FAs, 18:3Δ5,9,12—pinolenic acid, 20:3Δ5,11,14—sciadonic acid and 20:4Δ5,11,14,17—juniperonic acid. *—from flowering (March–April) until the end of red arils maturation (September–October).

Except Warsaw location, RAs proved to be a substantial source of linoleic (n-6 PUFA; 429–757 mg/100 g RA). High blood levels of n-6 acids were considered an increased risk of inflammatory and allergic conditions in epidemiological studies [48]. An increased intake of α-linolenic from the diet has the potential to limit the production of n-6 derived proinflammatory mediators and to enhance the biological efficacy of long chain n-3 PUFA [50]. As shown in Table 6, samples of RAs from two sites were an especially valuable source α-linolenic FA; 100 g of fruits may provide 832 to 938 mg of 18:3Δ9,12,15 acid. It should also be pointed out that the beneficial ratio of n-6/n-3 PUFAs in RAs is 1:0.8–1:1.7, as shown in Table 6. This ratio was much lower compared with the current ratio n-6/n-3 PUFAs of Western diet, from 15:1 to 16.7:1 [49].

Based on these results, RAs may represent a novel vegetable dietary source of valuable PUFAs belonging to n-3 family, including the long-chain ones and also unique PMI-FAs.

2.5. Elemental Characteristic

The contents of macroelements (K, P, S, Ca, Mg, Na), microelements (Zn, Fe, B, Cu, Mn, Cr, Mo, Co), and metals (Al, Ni, Bi, Ba, In, Ti, Li, Ag, Cd, Ga), are presented in Table 7. Most of them were significantly ($p < 0.05$) dependent on the site of RA collection.

Potassium (K) was the most abundant element in RA (772–878 mg/100 g), followed by P, S, Ca, and Mg. It was about 2 to 3-fold higher than reported for five species of wild fruits in Saudia Arabia and mulberries in China [2,9]. A much higher amount of potassium than in RAs was found in goji berries (2100 mg/100 g) [43]. RAs had potassium content comparable to that of bananas, regarded as a typical potassium source in the diet [51]. Potassium is an essential mineral, important to maintain body water and to participate in transmitting nerve impulses to muscles [51]. An adult human being needs approximately 4700 mg K per day, thus, RAs consumption can meet the daily required amount [52].

RAs contained, on average, about 100 mg/100 g phosphorus (P), which does not meet the recommended daily allowance (RDA) for adults [52]. In addition, the content of sodium (Na) was low (0.86 to 4.90 mg/100 g). However, it should be emphasized that P and Na consumption in developed countries exceeds RDA mainly due to the nearly ubiquitous distribution of phosphorus-based food additives [53].

Table 7. Elements composition of red arils (mean $\pm$ SE, $n = 9$).

Element	Fruit Collection Site			
	Zielona Góra	Warsaw	Koszalin	Cracow
	Macroelements (mg/100 g of fresh weight)			
K	772.29 $\pm$ 26.27 [a]	878.38 $\pm$ 5.14 [b]	838.18 $\pm$ 41.19 [a]	789.62 $\pm$ 50.86 [a]
P	95.96 $\pm$ 1.76 [a]	109.35 $\pm$ 6.14 [a]	101.03 $\pm$ 6.24 [a]	97.89 $\pm$ 7.08 [a]
S	29.13 $\pm$ 1.38 [a,b]	30.99 $\pm$ 2.31 [b]	27.78 $\pm$ 1.07 [a,b]	24.97 $\pm$ 0.52 [a]
Ca	20.75 $\pm$ 0.49 [a]	21.05 $\pm$ 0.53 [a]	23.69 $\pm$ 0.09 [b]	19.88 $\pm$ 0.33 [a]
Mg	19.53 $\pm$ 0.28 [a]	25.71 $\pm$ 0.58 [c]	22.88 $\pm$ 0.47 [b]	24.63 $\pm$ 0.34 [c]
Na	1.12 $\pm$ 0.20 [a]	4.90 $\pm$ 1.26 [c]	2.81 $\pm$ 0.20 [b]	0.86 $\pm$ 0.39 [a]
	Microelements (µg/100 g of fresh weight)			
Zn	1506.54 $\pm$ 16.98 [b]	947.72 $\pm$ 32.46 [a]	1146.59 $\pm$ 89.23 [a]	1080.24 $\pm$ 56.71 [a]
Fe	1111.39 $\pm$ 28.67 [a]	1447.57 $\pm$ 80.42 [b]	2537.13 $\pm$ 142.95 [c]	976.19 $\pm$ 68.97 [a]
B	619.40 $\pm$ 2.61 [b]	1152.07 $\pm$ 11.10 [c]	463.87 $\pm$ 63.22 [a]	652.87 $\pm$ 50.63 [b]
Cu	240.53 $\pm$ 1.24 [b]	251.23 $\pm$ 24.40 [a]	206.10 $\pm$ 19.21 [a]	201.84 $\pm$ 16.15 [a]
Mn	103.87 $\pm$ 1.93 [b]	76.45 $\pm$ 2.62 [a]	521.09 $\pm$ 5.35 [c]	721.55 $\pm$ 13.56 [d]
Cr	15.42 $\pm$ 0.64 [b,c]	16.36 $\pm$ 1.60 [c]	13.57 $\pm$ 1.16 [b,c]	11.18 $\pm$ 0.53 [a]
Mo	11.06 $\pm$ 0.43 [a]	17.68 $\pm$ 0.28 [b]	11.43 $\pm$ 0.03 [a]	17.85 $\pm$ 0.48 [b]
Co	7.34 $\pm$ 0.21 [b]	0.13 $\pm$ 0.03 [a]	7.53 $\pm$ 0.09 [b]	6.03 $\pm$ 0.17 [b]
	Metals (µg/100 g of fresh weight)			
Al	466.48 $\pm$ 24.80 [b]	616.56 $\pm$ 17.23 [c]	1855.06 $\pm$ 148.40 [d]	306.63 $\pm$ 7.16 [a]
Ni	129.52 $\pm$ 1.68 [a]	98.93 $\pm$ 8.02 [a]	409.77 $\pm$ 4.23 [c]	217.53 $\pm$ 4.87 [b]
Bi	62.65 $\pm$ 0.29 [b,c]	54.15 $\pm$ 14.44 [b]	46.02 $\pm$ 21.56 [b]	22.17 $\pm$ 17.43 [a]
Ba	37.21 $\pm$ 1.23 [b]	34.18 $\pm$ 1.79 [b]	77.64 $\pm$ 12.40 [c]	24.63 $\pm$ 0.41 [a]
In	33.40 $\pm$ 2.48 [b]	23.92 $\pm$ 3.18 [a]	19.24 $\pm$ 2.84 [a]	22.52 $\pm$ 2.79 [a]
Ti	32.55 $\pm$ 3.35 [b]	47.67 $\pm$ 1.03 [c]	71.12 $\pm$ 3.81 [d]	17.03 $\pm$ 2.26 [a]
Li	13.40 $\pm$ 0.53 [c]	22.02 $\pm$ 0.46 [d]	10.42 $\pm$ 0.58 [b]	3.16 $\pm$ 0.29 [a]
Ag	14.32 $\pm$ 1.50 [a]	12.85 $\pm$ 0.93 [a]	11.80 $\pm$ 0.65 [a]	12.17 $\pm$ 0.54 [a]
Cd	7.41 $\pm$ 2.46 [b]	9.21 $\pm$ 1.09 [b]	4.78 $\pm$ 0.52 [a]	24.66 $\pm$ 3.48 [c]
Ga	7.29 $\pm$ 0.34 [b]	11.62 $\pm$ 0.90 [c]	16.21 $\pm$ 1.14 [d]	2.40 $\pm$ 0.61 [a]

[a, b c, d]—values bearing the same superscripts in rows do not differ significantly ($p < 0.05$) from each other.

The content of calcium (Ca) in RAs (about 20 mg/100g) was much higher than in blackberries, red raspberries, strawberries, and cherries [35]. However, mulberries and goji berries have more reach in calcium than RAs (71–124 g/100 g and 126–149 mg/100 g, respectively) [2,43].

Despite the physiologic role of magnesium (Mg) and its proven or potential benefits, its dietary intake is known to be inadequate in many populations [54]. The abundance of magnesium in RAs (23 mg Mg/100 g, on average) allows meeting only 7% of its RDA for a healthy adult, as in the case of mulberries [2], while the consumption of goji berries contributed to 15% of RDA of magnesium [43].

With regard to microelements, irrespectively of the collection site, RAs had a substantial amount of zinc (Zn; 948–1507 µg/100g), similar to the wild bilberries from the Eastern Italian Alps, but much more than many cultivated berries [3,34,43]. Zinc is necessary for many enzymatic reactions and for the absorption of B-group vitamins. It allows maintaining healthy skin, self-immunity, and good functioning of the prostate gland [54]. Consumption of a 100 g portion of RAs allows meeting from 11 to 15% of the RDA of Zn for adults, depending on sex.

Samples of RAs differed significantly ($p < 0.05$) in iron (Fe) content (976–2537 µ/100 g) depending on the collection site (Table 2). Iron, as the constituent of hemoglobin, myoglobin, and of many enzymes, is an essential nutrient. Its adequate supply is especially important for females aged 14 to 50 years. Consumption of a 100 g portion of RAs allows meeting from 9 to 15% of the RDA for adults. However, according to literature, 100 g serving of other berries (blueberry, blackberry and goji berry) cover a higher contribution of RDA of iron (21 to 90%) than in the case of RAs [35,43].

Manganese (Mn) is required for macronutrient metabolism. No formal RDA for Mn was established, but the US NRC set an estimated safe and adequate dietary intake of 2 two 5 mg/day for adults [55]. RAs differed ($p < 0.05$) in Mn content between collection sites: much higher levels were noted in samples from Koszalin and Cracow sites (521–722 µ/100 g) than from Warsaw and Zielona Góra sites (76.5–103.9 µ/100 g). Generally, as compared with literature data about Mn content in other fruits (wild strawberry-tree fruits), RAs may be regarded as a good source of Mn [36]. However, as compared with RA, much higher Mn content was found in wild bilberries, most likely because of the specific composition of soil in the natural habitat of the Eastern Italian Alps [3]. In addition, goji berries, organically grown using organic fertilizers, were rich in Mn (980 g/100 g) [43].

The content of copper in RAs (225 µg/100 g, on average) was about twice higher than in mulberries, but much lower than in goji berries and *Rosa sterilis* fruits [2,8,43].

Boron (B) was the third quantitatively important element in RAs (464–1152 µg/100 g) and proved to be a rich source of boron, similar to many popular nuts (Brazil, pistachio, cashew) [56]. Boron plays an important role in osteogenesis, its deficiency was shown to adversely impact bone development and regeneration, and to support the effects of estrogen, testosterone, and vitamin D. It was suggested that humans need at least 0.2 mg/d of boron, and that the diet should provide approximately 1 to 2 mg B/d [56].

The presented results are thought to reflect soil types and properties of natural sampling sites, as some elements (Fe, Mn, Cu) are abundant in podsols and brown acid soils [3,8]. Micronutrients may vary largely depending on environmental conditions, such as precipitation, humidity and soil composition, as they could induce responses to physiological stress, when the minerals could act as cofactors regulating the metabolic pathways of the plant [36].

The contents of chromium (Cr), molybdenum (Mo), and cobalt (Co) were relatively low (Table 7). The minimum RDA for Cr amounts to 24 µg/day for most adults [57]. Consumption of a 100 g portion of RAs meets 30 to 50% of its RDA. Shim fruits (*R. tomentosa*) grown in Vietnam is a better source of Cr than RAs [40]. Cobalt is included in vitamin B_{12}-cobalamine and plays a very important role in the synthesis of AAs and some proteins in nerve cells, and in producing neurotransmitters. The RDA of Co is 5 µg/day. The content of Co in a 100 g portion of RAs exceeded the RDA, except for the samples from Koszalin site (Table 7).

Regarding the presence of metals, the content of aluminum (Al) dominated among the detected metals in RAs (Table 7). Anthropogenic sources of many metals in soils are natural processes (e.g., weathering of rocks), mining and smelting activities, use of sewage sludge and phosphate fertilizers, which may contain heavy metals as impurities. It should be pointed out that the accumulation of trace metals is a normal and essential process for the growth and nurturing of plants [57]. Thus, RAs grown at the Koszalin site on brown acid soil contained 3- to 6-fold more aluminum (Al) than RAs from other sites (Table 7) [3]. The Al content in e.g., bilberries, lingonberries, and rosehips in Finland was higher than in RAs [58]. Since Al is toxic, the EFSA established a Tolerable Weekly Intake (TWI) of 1 mg Al per kg of body weight [59].

RAs from the Koszalin site contained a 2- to 4-fold higher amount of nickel (Ni) compared with samples from other sites, most likely due to environment contamination [57]. Contents of Ni in RAs collected from other sites were similar to that found in many fruits [40,57,58].

Cadmium (Cd) is a heavy metal, especially toxic for kidneys, but may also induce bone demineralization, and was classified as carcinogenic to humans. Cereals and cereal products, vegetables, nuts and pulses, starchy roots and potatoes as well as meat and meat products, contribute most to human exposure. The EFSA has set TWI for cadmium at 2.5 µg Cd per kg of body weight (µg/kg BW) [60]. Given the health effects of cadmium on humans, its maximum level in fruits and vegetables was set at 0.05 mg/kg, accordingly EC 1881/2006 [61]. Depending on the site of fruit collection, RAs contained 4.78 to 24.66 µg Cd/100 g. The samples from Cracow site exceeded the acceptable Cd level almost five times

(Table 7), whereas in the samples from Koszalin site, its content was below the acceptable level, similar to freeze-dried strawberries in China [1]. In addition, it is noteworthy that the contents of other trace metals (Bi, Ba, In, Ti, Li, Ag, Ga) were below 80 μg/100 g, i.e., under their maximum permissible limits [3,62]. No Pb was detected in the RAs samples.

The presence of trace metals in fruits may be attributed not only to the natural background of heavy metal content in the soil geochemistry but also derived from the environmental pollution [40,57]. Thus, the growing environment and in particular the soil aluminum, cadmium and nickel concentration should therefore be taken into account when choosing harvest region [40].

3. Material and Methods

3.1. Sample Collection

Red berries of *Taxus baccata* L. were collected from plants growing in natural habitats in four different sites of Poland (West, Central, North and South), in the neighborhood of cities: Zielona Góra, Warsaw, Koszalin, and Cracow, respectively (Figure 2). In each site, red berries were harvested thrice (throughout September to October, 2018), from 10 trees each (from different parts of crown), growing in three places ($n = 9$). The plants were identified as *Taxus baccata* L. by morphologic comparisons of leaves, flowers, buds, bark of trees and berries, according to Seneta [63] and Krüssmann [64]. Soil and climate conditions at fruit collection sites are presented in Table 2 and Figure 1. Fruits were manually separated from the seeds to obtain RAs for analyses.

Figure 2. Location of collection of Taxus baccata red berries in natural habitats in Poland.

3.2. Analysis of Taxus Compounds

Sample preparation: aril samples (250 mg each) were dried for 24 h at 60 °C, mixed with 600 μL ammonium buffer (pH = 9), atropine-D3 (Sigma-Aldrich, Steinheim, Germany) as the internal standard, and 1.2 mL of dichloromethane, followed by vortexing for 2 min and centrifuged for 5 min. Next, the organic phase was separated and evaporated to dryness under a stream of nitrogen. The residues were then redissolved in mobile phase (water/acetonitrile, 90:10 v/v), and 20 μL of the eluate was injected into a HPLC column.

The separation was performed using a Luna Pentafluorophenyl (2) 100 A column (150 × 2 mm, 5 μm; Phenomenex, Aschaffenburg, Germany). LC-MS/MS analysis was conducted on an Agilent 1260 Infinity HPLC system (Agilent Technologies, Santa Clara, USA) coupled to a 3200 QTrap (AB Sciex, Darmstadt, Germany) equipped with an electrospray ionization (ESI) source. The separation was carried out with water/ammonium format (2 mM)/formic acid (0.2%) mixture (solvent A) and an acetonitrile/ammonium format (2 mM)/formic acid (0.2%) mixture (solvent B). The initial solvent ratio was 90:10 (A:B) and was gradually decreased to 10:90 (A:B) within 10 min, the flow rate being 0.5 mL/min. This was held for 5 min and a flowrate of 1 mL/min was applied. Then, the gradient went from 10:90 (A:B) at 15 min back to 90:10 (A:B) at 15.5 min using a flow rate of 1.5 mL/min; the temperature was 20 °C.

The MS source temperature was set to 630 °C, curtain gas to 35 psi, ion source gas 1 to 45, ion source gas 2 to 90 psi, collision gas (CAD) to medium and ion spray voltage to 5500 V. Individual compounds were detected in ESI+ mode and identified by multiple-reaction monitoring (MRM) mode following two mass transitions per analyte [18].

To mimic the red arils matrix, a calibration curve with redcurrant (*Ribes rubrum*) berries was used for quantification; 250 mg of redcurrant berries (dried for 24 h at 60 °C) were spiked with a mix of the five standards (Taxol A, 10-DAB III, BAC III, Cephalomannine and Taxinine M; Sigma Aldrich, Steinheim, Germany) in concentrations ranging from 0.002 to 40 μg taxanes per g, extracted and analyzed as described above.

3.3. Analysis of Proximate Composition

Dry matter, ash, and protein contents were determined according to AOAC procedures [65]. Ash content was determined by sample incineration in a muffle furnace (Nabertherm, Germany) at 550 °C. The extraction and determination of lipids from RAs were carried out using the Folch's method with chloroform-methanol mixture (2:1, *v/v*). The total energy content was computed as follows:

$$\text{Energy (kcal)} = 4 \times (\text{g protein} + \text{g carbohydrate}) + 9 \times (\text{g lipid}). \tag{1}$$

3.4. Analysis of Sugars

The RAs were homogenized for one minute (13,500 rpm) in 80% aqueous ethanol, using a DI 25 homogenizer (Ika Warke, Dusseldorf, Germany), and then centrifuged at $2490 \times g$ for 20 min in an MPW-260R device (Warsaw, Poland).

An HPLC analysis of sugars (glucose, fructose, sucrose) was performed using a Dionex Ultimate 3000 instrument (Thermo Scientific, Germany) equipped with a refractive index detector (RefractoMax 521). Separation of sugars was conducted on a LiChrospher 100-10 NH$_2$ (5 μ) column (250 × 4 mm). The isocratic elution mobile phase was provided using acetonitrile/water (87:13 *v/v*) at a flow rate of 1.3 mL/min. The identification of sugars was made by comparing the relative retention times of sample peaks with standards (Sigma Aldrich, Poland). Quantification of sugars was made using four-point calibration curves (in a concentration range of 0.1 to 1 mg/mL, for each compound). The contents of sugars were expressed as g/100g of fresh weight of RAs.

3.5. Analysis of Amino Acid Profile

Amino acids were quantified by HPLC after an acidic hydrolysis according to Dhillon, Kumar, and Gujar [66] and using an AccQ-Tag reagent kit from Waters (Milford, MA, USA) for derivatization of amino acids.

Each RAs sample (ca. 30 mg) was hydrolyzed with 4 mL of HCl and 15 μL of phenol at 110 °C for 24 h, and then entrapped in N$_2$ atmosphere. The hydrolyzate was filtered through syringe filters (0.45 μm), and then dried with N$_2$. Next, 10 μL of the sample was mixed with 70 μL of borate buffer (pH 8.2–9.0), then 20 μL of 6-aminoquinolyl-*N*-hydroxysuccinimidylcarbamate acetonitrile solution (3 mg/mL) were added to the mixture (AccQ-Tag reagent kit, Waters, Milford, USA). Analogous procedures were used in the case of standards.

The amino acid profile (AAs) was identified on a Dionex Ultimate 3000 HPLC instrument (Thermo Scientific, Germering, Germany). Separation was provided on a reverse-phase Nova-Pak C18 column (4 μm, 150 × 3.9 mm) (Waters, Milford, MA, USA) at 37 °C. Elution was run in a two-component gradient at 1 mL/min; eluent A: acetic-phosphate buffer, eluent B: acetonitrile-water (60:40). The detector (VWD-3400RS) was set at 240 nm wavelength. The AAs peaks were computed from AAs standards (Sigma-Aldrich, Poznan, Poland) run at five concentrations. Individual AA values were expressed as mg/100 g of fresh weight of RAs.

3.6. Analysis of Fatty Acid Composition

Lipid fraction extracted from each RAs sample was used for derivatization of triacylo-glycerols into methyl esters of fatty acids (FAMEs) for gas chromatography analysis (GC). Lipids were saponified by boiling in 0.5 mol/L NaOH solution for 10 min. The FAMEs were prepared by transmethylation using a catalyst (95% H_2SO_4). Briefly, the samples were heated in a water bath at 100 °C for 40 min in a mixture of sulfuric acid and methanol, followed by the addition of n-hexane. After cooling, saturated NaCl solution was added and mixed thoroughly. Finally, 1 µL of the upper phase containing FAMEs was injected into the chromatograph (GC) injection port.

The FAMEs were analyzed by GC using an Agilent 6890N (HP Agilent, Santa Clara, CA, USA) instrument equipped with a flame ionization detector, a capillary column with the stationary phase of high polarity (100 m, 0.25 mm I.D., film thickness 0.1 µm; Rtx 2330 Restek). The analyses involved a programmed run with temperature ramps. The initial oven temperature was 120 °C for 40 min, and was then ramped to 155 °C at 1.5 °C/min and held for 50 min. The temperature was then ramped again at 2 °C/min to 210 °C and held for 35 min. Injector and detector temperatures were maintained at 250 °C; hydrogen was used as the carrier gas at the flow rate of 0.9 mL/min. The peaks were identified by comparison with Supelco 37 No. 47885-U standards and PUFA standards (Sigma Aldrich, Poznan, Poland). Identification of peaks of polymethylene-interrupted FAs was achieved by using chromatograms FAMEs of lipid extracted from *Taxus baccata* seeds (presented in Figure S1) and accordingly published chromatogram [20]. The contents of individual FAs were expressed in g/100 g FAs.

3.7. Elemental Analysis

The lyophilized samples of RAs were ground (Fritsch Pulverisette 14, Germany) and microwave-mineralized in a CEM MARS-5 Xpress mineralizer (CEM World Headquarters, Matthews, NC, USA) in HNO_3 (65%). The contents of macroelements, microelements, and trace metals were determined by the inductively coupled plasma optical emission spectroscopy (ICP-OES) according to Pasławski and Migaszewski [67], using a high-dispersion ICP-OES (Prodigy Teledyne, Leeman Labs, New Hampshire, MA, USA).

3.8. Statistical Analysis

All analyses were performed in triplicate. The results were expressed as means and standard errors (SE) and subjected to one-way ANOVA followed by Tukey's test. Differences between mean values were considered significant at $p < 0.05$. Analyses were performed with Statistica 3.1 software (Statsoft, Inc.,Tulsa, OK, USA).

4. Conclusions

Red arils of *Taxus baccata* L. could be used as a vegetable source of high-quality protein with predominating shares of lysine and leucine and can serve as a branched-chain amino acid-rich new ingredient of human diet. Because of the low content of simple sugars, red arils can also be recommend as a low-calorie snack. In vegetarian diet, red arils may be regarded as a source of iron and zinc, providing 9 to 15% of the recommended daily allowance. Red arils of *Taxus baccata* may represent a novel dietary source of valuable PUFAs belonging to the n-3 family, and the unique polymethylene-interrupted FAs, such as pinolenic, sciadonic and juniperonic. In addition, the beneficial ratio of PUFAs n-6/n-3 (from 1:0.8 to 1:1.7), much lower than that in the Western die, t is to be noted. Depending on the location, the consumption of 100 g of red arils would provide 204 to 998 mg PUFAs of the n-3 family. It may thus be worth applying for the GRAS (Generally Recognized as Safe) status of red arils as a safe food additive.

Site differentiation in the contents of macronutrients, fatty acids, amino acids, and macro- and micro-elements in red arils resulted from different environmental conditions such as water availability (sum of precipitation), sunlight intensity, and soil parameters and composition.

Because of the tendency to accumulate toxic metals (Al, Ni, and Cd), red arils may be regarded as a valuable indicator of environmental contamination/pollution. Thus, the levels of toxic trace elements (Al, Ni, Cd) have to be determined before collecting fruits from natural habitats. Although the samples of red arils were free from taxine alkaloids, we recommend monitoring taxus compounds to ensure safety of consumers.

Full understanding of the nutraceutical potential of red arils requires a further systematic analysis of other fractions, e.g., of phenolic compounds and carotenoids, which is expected to be reported soon.

Supplementary Materials: Table S1: Comparison of fatty acid composition (g/100 g of FA) of red arils and seeds from the Koszalin site (mean ± SE). Figure S1: Gas chromatograms of fatty acid methyl esters prepared from lipid of seeds (A) and red arils (B) of *Taxus baccata*.

Author Contributions: Conceptualization: M.T., Ł.S., J.S., S.S., M.Ł., I.D., J.R., A.A.; investigation: M.T., Ł.S., J.S., S.S., M.Ł., I.D., J.R., A.A., D.B., J.P.; methodology: M.T., Ł.S., J.S., J.R., A.A., J.P.; writing—original draft: M.T., J.R., Ł.S., J.S., A.A.; writing—review and editing: J.R., M.T., A.A.; data curation: M.T., J.S., A.A., J.R.; visualization: A.A., M.T., J.R.; software: D.B., A.A., J.R. All authors have read and agreed to the published version of the manuscript.

Funding: This research was financed by the Ministry of Science and Higher Education of Poland.

Institutional Review Board Statement: Not applicable.

Informed Consent Statement: Not applicable.

Data Availability Statement: The data presented in this study are available in supplementary material.

Conflicts of Interest: The authors declare that they have no known competing financial interests or personal relationships that could have appeared to influence the work reported in this paper.

Sample Availability: Samples of the compounds are not available from the authors.

References

1. Hua, Z.; Wang, Z.Y.; Yang, X.; Zhao, H.T.; Zhang, Y.C.; Dong, A.J.; Jing, J.; Wang, J. Determination of free amino acids and 18 elements in freeze-dried strawberry and blueberry fruit using an Amino Acid Analyzer and ICP-MS with micro-wave digestion. *Food Chem.* **2014**, *147*, 189–194.
2. Jiang, Y.; Nie, W.J. Chemical properties in fruits of mulberry species from the Xinjiang province of China. *Food Chem.* **2015**, *174*, 460–466. [CrossRef] [PubMed]
3. Barizza, E.; Guzzo, F.; Fanton, P.; Lucchini, G.; Sacchi, A.G.; Lo Schiavo, F.; Nascimbene, J. Nutritional profile and productivity of bilberry (*Vaccinium myrtillus* L.) in different habitats of a protected area of the Eastern Italian Alps. *J. Food Sci.* **2013**, *78*, C673.
4. Jaakkola, M.; Korpelainen, V.; Hoppulab, K.; Virtanena, V. Chemical composition of ripe fruits of Rubus chamaemorus L. grown in different habitats. *J. Sci. Food Agric.* **2012**, *92*, 1324–1330. [CrossRef]
5. Zheng, G.Q.; Zheng, Z.Y.; Xu, X.; Hu, Z.H. Variation in fruit sugar composition of Lycium barbarum L. and Lycium chinense Mill. of different regions and varieties. *Biochem. Syst. Ecol.* **2010**, *38*, 275–284. [CrossRef]
6. Barros, L.; Carvalho, A.M.; Morais, J.S.; Ferreira, I.C.F.R. Strawberry-tree, blackthorn and rose fruits: Detailed characterization in nutrients and phytochemicals with antioxidant properties. *Food Chem.* **2010**, *120*, 247–254. [CrossRef]
7. Guo, M.; Shi, T.; Duan, Y.; Zhu, J.; Li, J.; Cao, Y. Investigation of amino acids in wolfberry fruit (*Lycium barbarum*) by solid-phase extraction and liquid chromatography with precolumn derivatization. *J. Food Compos. Anal.* **2015**, *42*, 84–90. [CrossRef]
8. He, J.Y.; Zhang, Y.H.; Ma, N.; Zhang, X.L.; Liu, M.H.; Fu, W.M. Comparative analysis multiple ingredients in *Rosa roxburghii* and *R. sterilis* fruits and their antioxidant activities. *J. Funct. Foods.* **2016**, *27*, 29–41. [CrossRef]
9. Hegazy, A.K.; Mohamed, A.A.; Ali, S.I.; Alghamdi, N.M.; Abdel-Rahman, A.M.; Al-Sobeai, S. Chemical ingredients and antioxidant activities of underutilized wild fruits. *Heliyon* **2019**, *5*, e01874. [CrossRef]
10. Thomas, P.A.; Polwart, A. Taxus baccata L. *J. Ecol.* **2003**, *91*, 489–524. [CrossRef]
11. Milutinović, M.G.; Stanković, M.S.; Cvetković, D.M.; Topuzović, M.D.; Mihailović, V.B.; Marković, S.D. Antioxidant and anticancer properties of leaves and seed cones from European yew (*Taxus baccata* L.). *Arch. Biol. Sci. Belgrade* **2015**, *67*, 525–534. [CrossRef]
12. Nawrocka-Grześkowiak, U.; Skuza, L.; Szućko, I. Genetic variability of yew in the selected Polish nature reserves. *Sylwan* **2015**, *159*, 37–44.
13. Wilson, C.R.; Sauer, J.M.; Hooser, S.B. Taxines: A review of the mechanism and toxicity of yew (*Taxus* spp.) alkaloids. *Toxicon* **2001**, *39*, 179–185. [CrossRef]
14. Glowniak, K.; Mroczek, T.; Zobel, A.M. Seasonal changes in concentrations of four toxoids in *Taxus baccata* L. during autumn-spring period. *Phytomedicine* **1999**, *6*, 135–140. [CrossRef]

15. Grobosh, T.; Schwarze, B.; Felgenhauer, N.; Riesselmann, B.; Roscher, S.; Binscheck, T. Eight cases of fatal and non-fatal poisoning with *Taxus baccata*. *Forensic Sci. Int.* **2013**, *227*, 118–126. [CrossRef]

16. Merill, C.; Miller, M.C., III; Johnson, K.R.; Willingham, M.C.; Fan, W. Apoptotic cell death induced by baccatin III, a precursor of paclitaxel, may occur without G2/M arrest. *Cancer Chemoter. Pharmacol.* **1999**, *44*, 444–452.

17. Kucukboyaci, N.; Sener, B. Biological activities of lignans from *Taxus baccata* L. growing in Turkey. *J. Med. Plant Res.* **2019**, *4*, 1136–1140.

18. Siegle, L.; Pietsch, J. Taxus ingredients in the red arils of *Taxus baccata* L. determined by HPLC-MS/MS. *Phytochem. Anal.* **2018**, *29*, 446–451. [CrossRef]

19. Tabaszewska, M.; Słupski, J.; Gębczyński, P.; Pogoń, K.; Skoczylas, Ł. Fiber content in aril yew (*Taxus baccata* L.) originates from different area of Poland. In Proceedings of the 13th International Conference on Polysaccharides-Glycoscience, Prague, Czech Republic, 8–10 November 2017; pp. 287–288.

20. Takagi, T.; Itabashi, Y. cis-5-Olefinic unusual fatty acids in seed lipids of gymnospermae and their dystibution in triacyloglycerols. *Lipids* **1982**, *17*, 716–723. [CrossRef]

21. Wolff, R.; Deluc, L.G.; Marpeau, A.M. Conifer seeds: Oil content and fatty acid composition. *J. Am. Oil Chem. Soc.* **1996**, *73*, 765–771. [CrossRef]

22. Wolff, R.L.; Pédrono, F.; Marpeau, A.M.; Christie, W.W.; Gunstone, F.D. The seed fatty acid composition and the distribution of Δ5-olefinic acids in the triacylglycerols of some taxaceae (*Taxus* and *Torreya*). *J. Am. Oil Chem. Soc.* **1998**, *75*, 1637–1641. [CrossRef]

23. Wolff, R.L.; Christie, W.W. Structures, practical sources (gymnosperm seeds), gas-liquid chromatographic data (equivalent chain lengths), and mass spectrometric characteristics of all-cis Δ5-olefinic acids. *Eur. J. Lipid Sci. Technol.* **2002**, 234–244. [CrossRef]

24. Morishige, J.; Amano, N.; Hirano, K.; Nishio, H.; Tanaka, T.; Satouchi, K. Inhibitory effect of juniperonic acid (Δ-5c, 11c, 14c, 17c - 20:4, Ω-3) on bombesin-induced proliferation of swiss 3T3 cells. *Biol. Pham. Bull.* **2008**, *31*, 1786–1789. [CrossRef] [PubMed]

25. Tsai, P.J.; Huang, W.C.; Lin, S.W.; Chen, S.N.; Shen, H.J.; Chang, H.; Chuang, L.T. Juniperonic acid incorporation into the phospholipids of murine macrophage cells modulates pro-inflammatory mediator production. *Inflammation* **2018**, *41*, 1200–1214. [CrossRef] [PubMed]

26. Chen, S.J.; Hsu, C.P.; Li, C.W.; Lu, J.H.; Chuang, L.T. Pinolenic acid inhibits human breast cancer MDA-MB-231 cell metastasis in vitro. *Food Chem.* **2011**, *126*, 1708–1715. [CrossRef] [PubMed]

27. Tili, N.; Tir, M.; Benlajnef, H.; Khemiri, S.; Mejri, H.; Rejeb, S.; Khaldi, A. Variation in protein and oil content and fatty acid composition of *Rhus tripartitum* fruits collected at different maturity stages in different locations. *Ind. Crops Prod.* **2014**, *59*, 197–201.

28. Ma, Y.; Tian, J.; Wang, X.; Huang, C.; Tian, M.; Wei, A. Fatty acid profiling and chemometric analyses for *Zantoxylum* pericarps from different geographic origin and genotype. *Foods* **2020**, *9*, 1676. [CrossRef]

29. Madawala, S.R.P.; Brunius, C.; Adholeya, A.; Tripathi, S.B.; Hanhineva, K.; Hajazimi, E.; Shi, L.; Dimberg, L.; Landberg, R. Impact of location on composition of selected phytochemicals in wild sea buckthorn (*Hippophae rhamnoides*). *J. Food Compos. Anal.* **2018**, *72*, 115–121. [CrossRef]

30. Fonseca, D.F.S.; Salvador, Â.C.; Santos, S.A.O.; Vilela, C.; Freire, C.S.R.; Silvestre, A.J.D.; Rocha, S.M. Bioactive phytochemicals from wild Arbutus unedo L. berries from different locations in Portugal: Quantification of lipophilic components. *Int. J. Mol. Sci.* **2015**, *16*, 14194–14209. [CrossRef]

31. Iszkuło, G.; Kosiński, P.; Hajnos, M. Sex influences the taxanes content in *Taxus baccata*. *Acta Physiol. Plant.* **2013**, *35*, 147–152. [CrossRef]

32. Veselá, D.; Šaman, D.; Valterová, I.; Vaněk, T. Seasonal variations in the content of taxanes in the bark of *Taxus baccata* L. *Phytochem. Anal.* **1999**, *10*, 319–321.

33. Sharma, H.; Garg, M. A review of traditional use, phytoconstituents and biological activities of Himalayan yew, *Taxus wallichiana*. *J. Integr. Med.* **2015**, *13*, 80–90. [CrossRef]

34. Gai, Q.Y.; Jiao, J.; Wang, X.; Liu, J.; Fu, Y.J.; Lu, Y.; Wang, Z.Y.; Xu, X.J. Simultaneous determination of taxoids and flavonoids in twigs and leaves of three Taxus species by UHPLC-MS/MS. *J. Pharmaceut. Biomed.* **2020**, *189*, 113456. [CrossRef]

35. De Souza, V.R.; Pereira, P.A.P.; Da Silva, T.L.T.; De Oliveira Lima, L.C.; Pio, R.; Queiroz, F. Determination of the bioactive compounds, antioxidant activity and chemical composition of Brazilian blackberry, red raspberry, strawberry, blueberry and sweet cherry fruits. *Food Chem.* **2014**, *156*, 362–368. [CrossRef]

36. Ruiz-Rodriguez, B.M.; Morales, P.; Fernández-Ruiz, V.; Sánchez-Mata, M.C.; Cámara, M.; Díez-Marqués, C.; Pardo-de-Santayana, M.; Molina, M.; Tardío, J. Valorization of wild strawberry-tree fruits (*Arbutus unedo* L.) through nutritional assessment and natural production data. *Food Res. Int.* **2011**, *44*, 1244–1253. [CrossRef]

37. Rutkowska, J.; Adamska, A.; Pielat, M.; Białek, M. Comparison of composition and properties of *Rosa rugosa* fruits preserved using conventional and freeze-drying methods. *Zywn-Nauk. Technol. Ja.* **2012**, *4*, 32–43. [CrossRef]

38. Cossignani, L.; Blasi, F.; Simonetti, M.S.; Montesano, D. Fatty acids and phytosterols to discriminate geographic origin of *Lycium barbarum* berry. *Food Anal. Methods* **2018**, *11*, 1180–1188. [CrossRef]

39. Bastos, C.; Barros, L.; Dueñas, M.; Calhelha, R.C.; Queiroz, M.J.R.; Santos-Buelga, C.; Ferreira, I.C. Chemical characterization and bioactive properties of *Prunus avium* L.: The widely studied fruits and the unexplored stems. *Food Chem.* **2015**, *173*, 1045–1053. [CrossRef]

40. Lai, T.N.H.; André, C.; Rogez, H.; Mignolet, E.; Nguyen, T.B.T.; Larondelle, Y. Nutritional composition and antioxidant properties of the sim fruit (*Rhodomyrtus tomentosa*). *Food Chem.* **2015**, *168*, 410–416. [CrossRef]
41. Shimomura, Y.; Kitaura, Y. Physiological and pathological roles of branched-chain amino acids in the regulation of protein and energy metabolism and neurological functions. *Pharmacol. Res.* **2018**, *133*, 215–217. [CrossRef]
42. Dulf, F.V. Fatty acids in berry lipids of six sea buckthorn (*Hippophae rhamnoides* L., *subspecies carpatica*) cultivars grown in Romania. *Chem. Cent. J.* **2012**, *6*, 106. [CrossRef] [PubMed]
43. Pedro, A.C.; Sánchez-Mata, M.C.; Pérez-Rodríguez, M.L.; Cámara, M.; López-Colón, J.L.; Bach, F.; Bellettini, M.; Haminiuk, I.C.W. Qualitative and nutritional comparison of goji berry fruits produced in organic and conventional systems. *Sci. Hortic.* **2019**, *257*, 108660. [CrossRef]
44. Issaoui, M.; Flamini, G.; Brahmi, F.; Dabbou, S.; Ben-Hassine, K.; Taamali, A.; Chehab, H.; Ellouz, M.; Zarrouk, M.; Hammamia, M. Effect of the growing area conditions on differentiation between Chemlali and Chétoui olive oils. *Food Chem.* **2010**, *119*, 220–225. [CrossRef]
45. Patil, M.M.; Muhammed, A.M.; Anu-Appaiah, K.A. Lipids and fatty acid profiling of major Indian *Garcinia* fruit: A comparative study and its nutritional impact. *J. Am. Oil Chem. Soc.* **2016**, *93*, 823–836. [CrossRef]
46. Zhou, X.; Shang, J.; Qin, M.; Wang, J.; Jiang, B.; Yang, H.; Zhang, Y. Fractionated antioxidant and anti-inflammatory kernel oil from *Torreya fargesii*. *Molecules* **2019**, *24*, 3402. [CrossRef]
47. Sánchez-Salcedo, E.M.; Sendra, E.; Carbonell-Barrachina, A.A.; Martinez, J.J.; Hernandez, F. Fatty acids composition of Spanish black (*Morus nigra* L.) and white (*Morus alba* L.) mulberries. *Food Chem.* **2016**, *190*, 566–571. [CrossRef]
48. Nowak-Dyjeta, K.; Giertych, M.J.; Thomas, P.; Iszkuło, G. Males and females of *Juniperus communis* L. and *Taxus baccata* L. show different seasonal patterns of nitrogen and carbon content in needles. *Acta Physiol. Plant.* **2017**, 191. [CrossRef]
49. Baker, E.J.; Miles, E.A.; Burdge, G.C.; Yaqoob, P.; Calder, P.C. Metabolism and functional effects of plant-derived omega-3 fatty acids in humans. *Prog. Lipid Res.* **2016**, *64*, 30–56. [CrossRef]
50. Goyens, P.L.; Spilker, M.E.; Zock, P.L.; Katan, M.B.; Mensink, R.P. Conversion of alpha-linolenic acid in humans is influenced by the absolute amounts of alpha-linolenic acid and linoleic acid in the diet and not by their ratio. *Am. J. Clin. Nutr.* **2006**, *84*, 44–53. [CrossRef]
51. Hardisson, A.; Rubio, C.; Baez, A.; Martin, M.; Alvarez, R.; Diaz, E. Mineral composition of banana (*Musa acuminate*) from the island of Tenerife. *Food Chem.* **2001**, *73*, 153–161. [CrossRef]
52. Jarosz, M. (Ed.) *Nutritional Standards for the Polish Population*; Institute of Food and Nutrition: Warsaw, Poland, 2017; pp. 238–256.
53. Carrigan, A.; Klinger, A.; Choquette, S.S.; Luzuriaga-McPherson, A.; Bell, E.K.; Darnell, B.; Gutiéerrez, O.M. Contribution of food additives to sodium and phosphorus content of diets rich in processed foods. *J. Ren. Nutr.* **2014**, *24*, 13–19. [CrossRef] [PubMed]
54. Heghedüs-Mindru, R.C.; Heghedüs-Mindru, G.; Negrea, P.; Sumalan, R.; Negrea, A.; Stef, D. The monitoring of mineral elements content in fruit purchased in supermarkets and food markets in Timisoara, Romania. *Ann. Agric. Environ. Med.* **2014**, *21*, 98–105.
55. Aschner, J.L.; Aschner, M. Nutritional aspects of manganese homeostasis. *Mol. Aspects Med.* **2005**, *26*, 353–362. [CrossRef] [PubMed]
56. Pizzorno, L. Nothing boring about Boron. *Integr. Med.* **2015**, *14*, 35–48.
57. Moodley, R.; Koorbanally, N.; Jonnalagadda, S.B. Elemental composition and fatty acid profile of the edible fruits of Amatungula (*Carissa macrocarpa*) and impact of soil quality on chemical characteristics. *Anal. Chim. Acta* **2012**, *730*, 33–41. [CrossRef] [PubMed]
58. Ekholm, P.; Reinivuo, H.; Mattila, P.; Pakkala, H.; Koponen, J.; Happonen, A.; Hellstroöm, J.; Ovaskainen, M.L. Changes in the mineral and trace element contents of cereals, fruits and vegetables in Finland. *J. Food Compos. Anal.* **2007**, *20*, 487–495. [CrossRef]
59. EFSA-European Food Safety Authority, Statement on the evaluation of a new study related to the bioavailability of aluminium in food. *EFSA J.* **2011**, *9*, 21–57.
60. EFSA-European Food Safety Authority, Cadmium in food Scientific opinion of the Panel on Contaminants in the Food Chain Scientific Opinion. *EFSA J.* **2009**, *980*, 1–139.
61. Commission Regulation (EC) No 1881/2006 of 19 December 2006 setting maximum levels for certain contaminants in foodstuffs. *Off. J. Eur. Comm.* **2006**, *364*, 5–24.
62. Rose, M.; Baxter, M.; Brereton, N.; Baskaran, C. Dietary exposure to metals and other elements in the 2006 UK Total Diet Study and some trends over the last 30 years. *Food Addit. Contam.* **2010**, *10*, 1380–1404. [CrossRef]
63. Seneta, W. *Conifer trees and shrubs, part II*, 2nd ed.; PWN: Warsaw, Poland, 1987; pp. 455–456.
64. Krüssmann, G. *Handbook of Conifers, 2nd Revised ed. With the Participation of Hans-Dieter Warda*; Paul Parey Publishing House: Berlin/Hamburg, Germany, 1983; pp. 318–319.
65. AOAC. *Official Methods of Analysis of the Association of Official Analytical Chemists*, 16th ed.; AOAC International: Arlington, VA, USA, 1995; Volume 2.
66. Dhillon, M.K.; Kumar, S.; Gujar, G.T. A common HPLC-PDA method for amino acid analysis in insects and plants. *Indian J. Exp. Biol.* **2014**, *52*, 73–79.
67. Pasławski, P.; Migaszewski, Z.M. The quality of element determinations implant materials by instrumental methods. *Pol. J. Environ. Stud.* **2006**, *15*, 154–164.

Article

New Bioactive Peptides Identified from a Tilapia Byproduct Hydrolysate Exerting Effects on DPP-IV Activity and Intestinal Hormones Regulation after Canine Gastrointestinal Simulated Digestion

Sandy Theysgeur [1,2], Benoit Cudennec [1,*], Barbara Deracinois [1], Claire Perrin [2], Isabelle Guiller [2], Anne Lepoudère [2], Christophe Flahaut [1,3] and Rozenn Ravallec [1,*]

1   UMR-T 1158, BioEcoAgro, University of Lille, F-59000 Lille, France; sandy.th@hotmail.fr (S.T.); barbara.deracinois@univ-lille.fr (B.D.); christophe.flahaut@univ-artois.fr (C.F.)
2   Diana Pet Food, F-56250 Elven, France; cperrin@diana-petfood.com (C.P.); iguiller@diana-petfood.com (I.G.); alepoudere@diana-petfood.com (A.L.)
3   UMR Transfrontalière BioEcoAgro N° 1158, University of Artois, F-62000 Arras, France
*   Correspondence: benoit.cudennec@univ-lille.fr (B.C.); rozenn.ravallec@univ-lille.fr (R.R.); Tel.: +33-(0)362268590 (B.C. & R.R.)

check for **updates**

Citation: Theysgeur, S.; Cudennec, B.; Deracinois, B.; Perrin, C.; Guiller, I.; Lepoudère, A.; Flahaut, C.; Ravallec, R. New Bioactive Peptides Identified from a Tilapia Byproduct Hydrolysate Exerting Effects on DPP-IV Activity and Intestinal Hormones Regulation after Canine Gastrointestinal Simulated Digestion. *Molecules* **2021**, *26*, 136. https://doi.org/10.3390/molecules26010136

Academic Editors: María Dolores Torres and Elena Falqué López
Received: 20 November 2020
Accepted: 23 December 2020
Published: 30 December 2020

Publisher's Note: MDPI stays neutral with regard to jurisdictional claims in published maps and institutional affiliations.

Abstract: Like their owners, dogs and cats are more and more affected by overweight and obesity-related problems and interest in functional pet foods is growing sharply. Through numerous studies, fish protein hydrolysates have proved their worth to prevent and manage obesity-related comorbidities like diabetes. In this work, a human in vitro static simulated gastrointestinal digestion model was adapted to the dog which allowed us to demonstrate the promising effects of a tilapia byproduct hydrolysate on the regulation of food intake and glucose metabolism. Promising effects on intestinal hormones secretion and dipeptidyl peptidase IV (DPP-IV) inhibitory activity were evidenced. We identify new bioactive peptides able to stimulate cholecystokinin (CCK) and glucagon-like peptide 1 (GLP-1) secretions, and to inhibit the DPP-IV activity after a transport study through a Caco-2 cell monolayer.

Keywords: bioactive peptides; in vitro gastrointestinal digestion; fish byproduct hydrolysate; cholecystokinin; glucagon-like peptide 1; DPP-IV inhibitory peptides

1. Introduction

The world population is projected to rise by 2 billion in the next 30 years, reaching 9.7 billion in 2050 [1]. This growth implies in an increase in food consumption, and the protein demand will significantly grow as a result of socio-economic changes, increased urbanization, rising incomes and the recognition of the role of protein in healthy diets. Dietary protein production exerts a high environmental impact, particularly for animal-derived protein, which causes high greenhouse gas emissions, land-use changes linked to an important terrestrial biodiversity loss and a high-water demand [2]. In this context, there is an important need to valorize and to better characterize dietary protein derived-byproducts to optimize their use and to answer the worldwide growing protein demand. For instance, in 2018, about 25% of the 178 million tons of global fish and shellfish production were lost or wasted [3]. The valorization processes of fish high-quality protein byproducts will partially address these issues by offering a renewable alternative, whilst creating added value in numerous domains such as in functional food or the pet food industry. The global pet food market was valued at USD 103.5 billion in 2016 in which the segment of healthcare and nutritional supplements shared 5%. In parallel, overweight and obesity and their associated chronic diseases such as type 2 diabetes mellitus (T2DM) are growing at a worrying rate around the globe. In 2016, 650 million adults were obese,

19

amongst 1.9 billion overweight persons. In 2019, the prevalence of T2DM was estimated at 417 million and the projection for 2045 is about 630 million [4].

Like humans, companion dogs and cats are affected by overweight and obesity co-morbidities such as diabetes and cancers, leading to impaired health and reduced life span. Depending on breeds and the methodology used to evaluate health status, overweight and obesity prevalence was estimated between 19.7% and 59.3% in dogs and between 7% to 63% in cats. This situation is mainly due to excessive food offer and related calorie intake, as pet owners do not follow nutritional guidance, and to a loss of physical exercise leading to the overweight-derived problems mentioned above but also to skin disorders, respiratory and locomotor diseases [5,6].

Dietary protein digestion produces the release of peptides and free amino acids, which regulate short term food intake. Protein-digested products stimulate the secretion of satiety signals via the "intestinal sensing" phenomenon, a nutrient recognition on the apical side of the enteroendocrine cells (EECs) [7,8]. Nevertheless, mechanisms that lead to gut hormones secretion by EECs after peptide and amino acid intestinal sensing remain unclear [9]. The two well-known intestinal anorexigenic hormones, cholecystokinin (CCK) and glucagon-like peptide-1 (GLP-1), exert their satiating effect via different pathways. GLP-1 also plays a significant role in glucose metabolism by regulating blood glucose via its incretin action [10]. After its secretion by EECs following a meal, circulating level of GLP-1 increases but has a short half-life because it is inactivated by the dipeptidyl peptidase 4 enzyme (DPP-IV) which is a serine protease present in a soluble form (plasma, urine, amniotic fluid) as well as in a membranous form in a wide type of cells and organs like the intestine, kidney or liver [11]. Hence, GLP-1 agonists and DPP-IV inhibitors have been targeted to treat insulin resistance occurring in T2DM [12]. These past few years, numerous dietary protein-derived peptides have been identified as DPP-IV inhibitors. Albeit they are less potent than drugs, they have shown a rising interest as a natural alternative to chemical DPP-IV inhibitors which harbor important side effects [13].

Fish protein- and hydrolysate-derived bioactive peptides have been identified to exert many in vitro and in vivo bioactivities, suggesting promising health benefits via several pathways involved in hypertension, obesity, inflammation, or in the regulation of glucose homeostasis in metabolic disorders [14]. Indeed, numerous studies have shown, in vitro as well as in vivo, the beneficial effects of fish hydrolysates on food intake regulation by the stimulation of the gut hormones secretion, in particular CCK and GLP-1 [15–18]. Moreover, fish hydrolysates could improve glucose homeostasis by increasing plasma GLP-1, gastric inhibitory peptide, also known as glucose-dependent insulinotropic polypeptide (GIP) increasing insulin secretion and lowering blood glucose [19,20] but also by reducing DPP-IV activity in vitro and in vivo [21–23].

The Nile tilapia (*Oreochromis niloticus*) is the third most produced species in the world for which the production was more than 4500 thousand tons representing 8.3% of the world aquaculture production in 2018. These days the development of fish processing has resulted in growing quantities of byproducts which can represent more than 70 percent of the processed fish [3].

In this work, we investigated whether a tilapia fish byproduct protein hydrolysate (FBPH) compared to its raw material (FBP), both submitted to an in vitro simulated gastrointestinal digestion (SGID), could stimulate gut hormones secretion in EECs and inhibit intestinal DPP-IV activity. This work aimed to future pet applications, so the consensual human SGID INFOGEST protocol was adapted to the dog digestion [24].

2. Results

2.1. Peptide Profile Modifications during SGID

To characterize and compare the impact of SGID on elution profiles and peptide apparent molecular weight (MW) distributions, FBPH and FBP, gastric and intestinal digests were submitted to SEC-FPLC. The oral digest both for FBPH and FBP serves as a referent peptide elution profile. The gastric and intestinal SGID phases did not extensively modify the shape of the peptide elution profiles of FBPH. In contrast, for FBP, significant modifications occurred during the SGID, as illustrated by the curve shift towards lower MW between the oral and the intestinal phases (Figure 1A). Besides, the MW distribution showed that the impact of the SGID was less significant for FBPH than for FBP. Thus, the proportions of high MW peptides (above 3 kDa) in oral, gastric and intestinal digests represented 72.8%, 66.6% and 48.1% for FBP and 52.4%, 50.9%, 45.4% for FBPH, respectively (Figure 1B). Moreover, in the intestinal phase, high MW peptides (above 6 kDa) disappeared entirely for both FBP and FBPH digests. The same phenomenon was observed for small MW peptides (below 1 kDa) for which the proportion increased during the SGID in a less extent manner for FBPH than for FBP. Indeed, their proportions in oral, gastric and intestinal digests were 27.5%, 27.5%, and 32.0% for FBPH and 19.0%, 19.0% and 32.9% for FBP, respectively. Despite slight differences, at the end of the SGID, the MW distribution profiles of FBP and FBPH were similar.

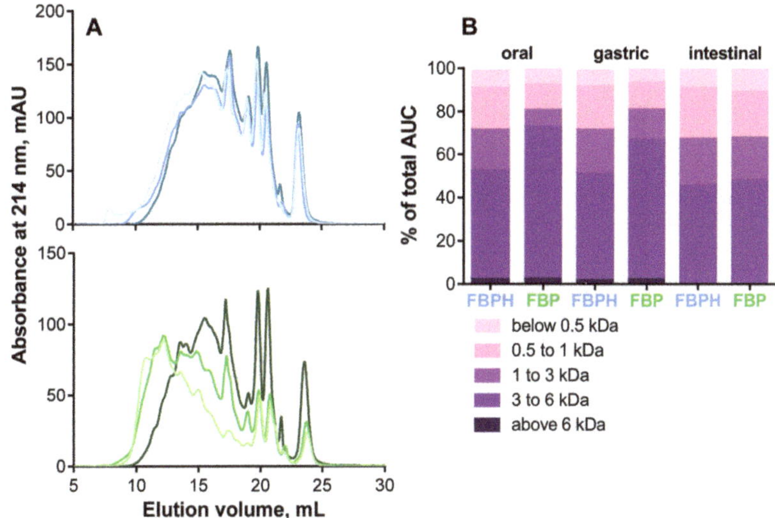

Figure 1. Peptide profiles and peptide molecular weight (MW) distributions. (**A**) Peptide profiles were obtained by size exclusion chromatography-fast protein liquid chromatography (SEC-FPLC): FBPH (blue curves), FBP (green curves); oral (light colored curves), gastric (colored curves) and intestinal (dark colored curves) digests. (**B**) The MW distribution of peptides in the different SGID compartments, expressed in percentage of total area under the curve (AUC), was calculated from the linear regression relationship which correlates the Log of known MW standard peptides and the elution volume.

2.2. CCK and GLP-1 Secretion Induced by FBPH and FBP Digests

The STC-1 cells exposure to increasing concentrations (2, 5 and 10 mg mL^{-1} w/v) of oral and gastric digests of FBP and FBPH induced a dose-dependent increase in CCK release (Figure 2A). At the higher concentration tested (10 mg mL^{-1} w/v), the FBPH contact led to a better stimulating effect with 7.2 ± 0.6 and 7.8 ± 0.2-fold of control (FOC) while the amounts of CCK obtained after FBP contact were 4.2 ± 0.4 and 5.7 ± 0.3 FOC for oral and gastric samples, respectively. Conversely, the intestinal digest of FBP highly stimulated the secretion of CCK (7.9 ± 0.7 FOC). The digestive process enhanced the ability of FBP to stimulate CCK secretion. Conversely, for FBPH, sole the intestinal phase of the SGID led to a slight diminution in CCK secretion in STC-1 cells at 10 mg mL (w/v).

The effects of oral and gastric digests of FBP and FBPH on GLP-1 secretion stimulation were equivalent (Figure 2B). Thus, only the gastric 10 mg mL (w/v) dose induced a significant increase of GLP-1 secretion with 14.6 ± 1.4 and 10.2 ± 2.3 FOC for FBP and

FBPH, respectively. After the intestinal phase, the stimulating effect of FBPH digest on GLP-1 secretion was highly enhanced and led to 20.6 ± 2.1, 43.3 ± 2.7 and 45 ± 3.0 FOC for 2, 5 and 10 mg mL (w/v) concentrations, respectively. The effect of the SGID intestinal phase on the ability of FBP to enhance the stimulation of GLP-1 secretion was weaker. Indeed, the results were significant only for 5 and 10 mg mL (w/v) doses with recovered GLP-1 secretions of 9.3 ± 1.3 and 29.4 ± 6.0 FOC, respectively.

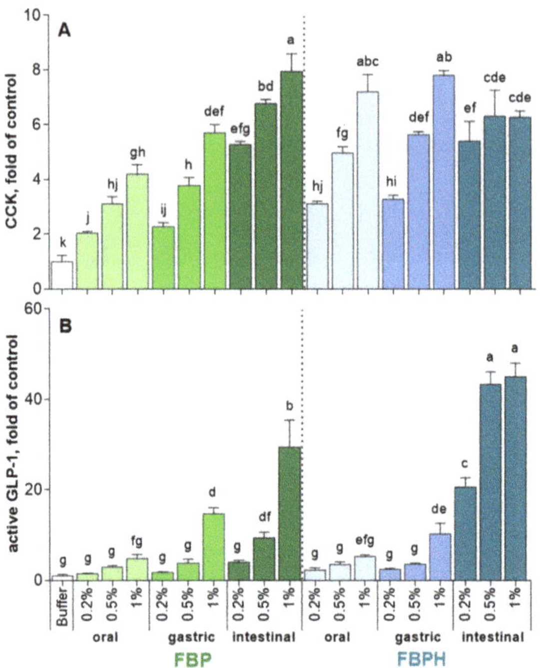

Figure 2. FBPH and FBP induction of intestinal hormones release during simulated GI digestion. The amounts of intestinal hormones released in STC-1 cells in the supernatants after 2 h of contact with the FBP and FBPH digests (2; 5 and 10 mg mL^{-1} w/v) were determined by radioimmunoassay for CCK (**A**) and active GLP-1 (**B**). Values are means $\pm$ SD and are expressed in fold of control (buffer). Means without a common letter within the same graph are significantly different ($p < 0.05$) using one-way ANOVA following by Tukey post-hoc test for pairwise comparisons.

2.3. Intestinal DPP-IV Inhibition Activity of FBPH and FBP Digests

The Caco-2 cells exposure to increasing concentrations (from 0.5 to 1.98 mg mL^{-1}, w/v) of oral, gastric and intestinal digests of FBP and FBPH induced a dose-dependent inhibition of the Caco-2 DPP-IV activity. The DPP-IV inhibitory activity potential of FBP increased through the different phases of the SGID. Indeed, the DPP-IV inhibitory activity observed with FBP digests assayed at 1.98 mg mL^{-1} (w/v) was 1.9-fold higher for intestinal digest than for oral sample. Moreover, the calculated IC$_{50}$ for the intestinal digest (IC$_{50}$ = 3.70 mg mL^{-1}) is about 23-fold lower than for the oral sample (IC$_{50}$ = 86.08 mg mL^{-1}) (Figure 3). For FBPH, the percentage of DPP-IV inhibitory activity of the samples collected in the three SGID compartments reached approximately 80% at 1.98 mg mL^{-1} (w/v). The calculated IC$_{50}$ was very closed for the oral, gastric and intestinal digests (insert of Figure 3). This highlights the small effect of the SGID on the DPP-IV inhibitory activity of FBPH. Results also showed that the intestinal digest of FBPH was much more potent than the FBP one. Thus, at a concentration of 1.98 mg mL^{-1} (w/v), the DPP-IV activity inhibition was about 2.3-fold higher for FBPH than for FBP with a 5.5-fold lower calculated IC$_{50}$ (Figure 3).

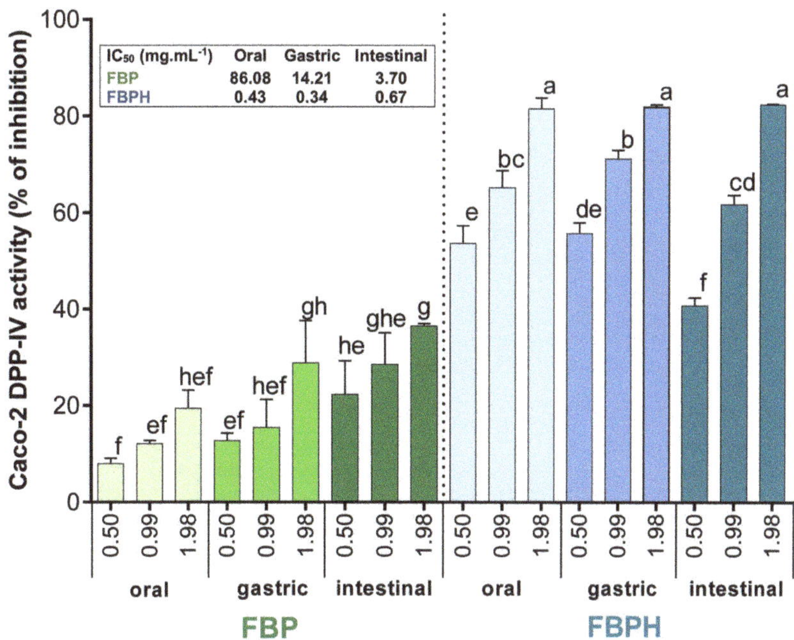

Figure 3. FBPH and FBP effects on the Caco-2 DPP-IV activity inhibition during SGID. The Caco-2 DPP-IV activity inhibition (%) obtained with FBP and FBPH digests assayed at increasing concentrations (0.5; 0.99 and 1.98 g L^{-1}, *w/v*). Means without a common letter within the same graph are significantly different ($p < 0.05$) using one-way ANOVA followed by Tukey post-hoc test for pairwise comparisons. Inset: IC_{50} were determined by linear regression correlating the DPP-IV activity inhibition percentage and the Ln of the concentration.

2.4. CCK and GLP-1 Secretion-Stimulating Peptides Identification

2.4.1. SEC and RP-HPLC Fractionation of FBPH Intestinal Digest

To identify active peptides able to stimulate the secretion of intestinal hormones, we first performed the SEC fractionation of the FBPH intestinal digest (Figure 4A). Four fractions were recovered and put in contact with STC-1 cells at 5 mg mL^{-1} (*w/v*) for 2 h. Results obtained showed that all of them were able to stimulate CCK secretion with the F2 and F4 fractions that displayed the higher potential with 3.5- and 4.5-fold of the control CCK secretion level, respectively (Figure 4B). Regarding GLP-1, the F2, F3 and F4 fractions were able to stimulate its secretion in STC-1 cells. The F2 fraction exerted a broadly higher potential than other fractions with 37 FOC and 2.6-fold of the FBPH digest (Figure 4C).

The F2 fraction was thus selected to be fractionated by RP-HPLC on a C18-column and 7 subfractions were designed (Figure 5A). The subfraction FE presented the higher potential to stimulate both CCK (Figure 5B) and GLP-1 (Figure 5C) secretion in STC-1 cells compared with other subfractions. Indeed, the CCK secretion stimulation by the FE subfraction was 31.9-, 8.4- and 7.2-fold higher than those obtained with the buffer, the F2 fraction and the FBPH intestinal digest, respectively. In the same way, the GLP-1 secretion stimulation by the FE subfraction was 32.0-, 2.5- and 17-fold higher than those obtained with the buffer, the F2 fraction and the FBPH intestinal digest, respectively.

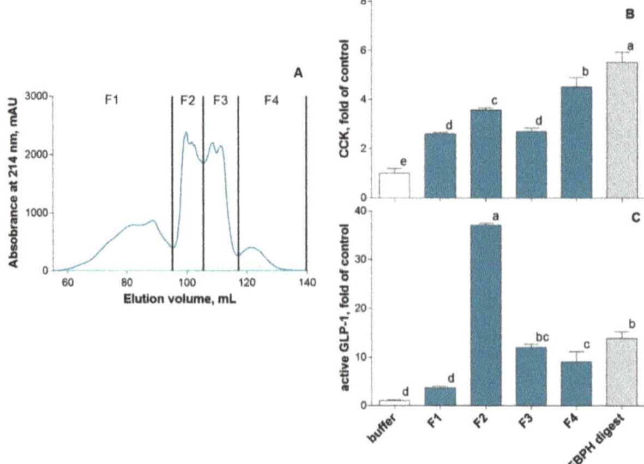

Figure 4. FBPH intestinal digest SEC fractionation effects on gut hormones release in STC-1 cells. The SEC fractionation of the FBPH intestinal digest was performed using a HiLoad 16/600 Superdex prepgrade column with an isocratic gradient of 30% acetonitrile 0.1% TFA (**A**). The amounts of intestinal hormones in the supernatants, after 2 h of contact with the fractions and or the FBPH digest (0.5% *w/v*), were determined by radioimmunoassay for CCK (**B**) and active GLP-1 (**C**). Values are the means of three repeated measurements and are expressed in fold of control (buffer) ± SD. Means without a common letter within the same graph are significantly different ($p < 0.05$) using one-way ANOVA followed by a Tukey post-hoc test for pairwise comparisons.

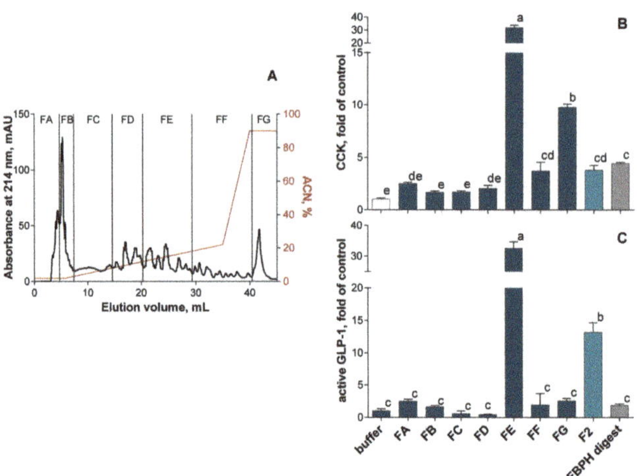

Figure 5. SEC-F2 RP-HPLC fractionation effect on gut hormones release in STC-1 cells. The RP-HPLC fractionation of the FBPH intestinal digest's SEC-F2 fraction was performed using a C18 Gemini column (150 × 10 mm, particles of 5 μm, 110 Å, Phenomenex) with an ACN gradient represented in red (**A**). The amounts of intestinal hormones released in the supernatants, after 2 h of contact, with the subfractions, the SEC-F2 fraction and the FBPH digest (0.5% *w/v*), were determined by radioimmunoassay for CCK (**B**) and active GLP-1 (**C**). Values are means of three repeated measurements and are expressed in fold of control (buffer) ± SD. Means without a common letter within the same graph are significantly different ($p < 0.05$) using one-way ANOVA following by Tukey post-hoc test for pairwise comparisons.

2.4.2. RP-HPLC-MS/MS Peptides Identification in the FE Subfraction

The FE subfraction was then subjected to RP-HPLC-MS/MS analysis to identify peptides present in this fraction. The Figure 6 showed the mass signal 3D-map obtained.

A total of 1739 peptide sequences were identified (database + de novo with ALC > 80%, data not shown). Among all the identified peptides, 20 of them were selected on the basis of (i) their presence in the most intense peaks of the UV chromatogram (λ = 214 nm), (ii) their ion intensity and (iii) their ion fragmentation quality (Table 1). Those peptides were then chemically synthesized and their ability to stimulate CCK and GLP-1 secretion in STC-1 cells was further assayed.

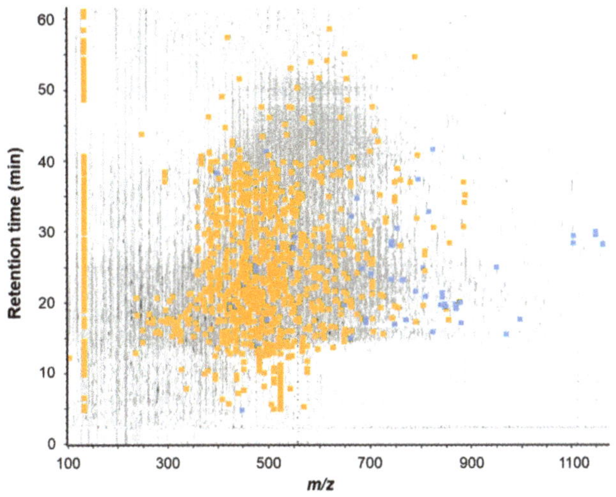

Figure 6. Mass signal 3D-map of the FE subfraction issued from the RP-HPLC separation of the FBPH intestinal digest's SEC-F2 fraction. Peptide map showing the repartition of all peptides detected by RP-UPLC-MS/MS analysis according to their retention time during chromatography, their mass to charge ratio (m/z) and their intensity. Grey signals represent all ions detected, blue squares represent identified peptides by database confrontation (false discovery rate (FDR) < 1%) and orange squares represent peptides sequenced by de novo mode (ALC score > 80%).

Table 1. List of the 20 peptides selected for chemical synthesis, following their identification by database confrontation or de novo sequencing. Peptides were listed according to their RP-UPLC-MS/MS characteristics (retention time (RT), mass to charge ratio (m/z) and their identification score displayed (i) by the average local confidence (ALC) score for the de novo sequencing, (ii) by the $-10^{\log P}$ score for database confrontation with the mass error (in ppm) for both identification modes (ID). nd: when the identification of the parent protein was not possible.

ID Mode	Sequence	RT	m/z	$10^{-\log P}$	ppm	Identification
	SAGPQGPIGPR	27.11	518.78	46.51	0.6	Collagen type I
	DVSGGYDE	20.69	841.33	41.49	6.2	Collagen type I
	HIHVNGA	16.85	747.39	38.12	−1.5	Collagen type 10
	VAPEEHPT	17.03	879.42	36.41	3.5	Alpha-actin
Database	AGPQGPIGPR	24.51	475.26	35.37	0.0	Collagen type I
	GATGPAGAV	23.75	700.36	32.09	0.2	Collagen type I Alpha-actin
	EAPLNPK	17.44	768.42	31	−4.6	Alpha-actin
	PEEHPT	16.74	709.32	27.74	8.3	Alpha-actin

Table 1. *Cont.*

ID Mode	Sequence	RT	*m/z*	ALC	ppm	
	LGVDE	21.54	532.26	98	2.6	nd
	VVEP	19.09	443.24	96	−10.4	nd
	LTDY	20.97	511.25	96	15.1	nd
	PSLVH	17.43	552.30	96	−33.1	nd
	ELLK	16.65	502.31	95	−31.5	nd
	LGME	21.28	449.21	95	6.9	nd
De novo	DLVDK	17.39	589.32	95	10.6	nd
	EVLSQ	21.84	575.29	95	−23.8	nd
	LKPT	20.48	235.67	92	−9.1	nd
	DSKPGSL	17.48	703.37	92	6.1	nd
	LLMMK	17.19	318.18	90	−26.9	nd
	LEL	18.83	374.23	80	−1.6	nd

2.4.3. CCK and GLP-1 Secretion Stimulation Induced by Synthetic Peptides from FBPH Intestinal Digest

The synthetized peptides were put in contact with STC-1 cells for 2 h at a final concentration of 1 mM and the amounts of secreted CCK and GLP-1 were further determined by radioimmunoassay. As shown in Figure 7A, among the 20 peptides assayed almost two of them, DLVDK and PSLVH, were able to significantly stimulate CCK secretion ($p < 0.0001$). Regarding GLP-1, only the LKPT peptide was able to significantly stimulate active-GLP-1 release ($p < 0.001$) (Figure 7B).

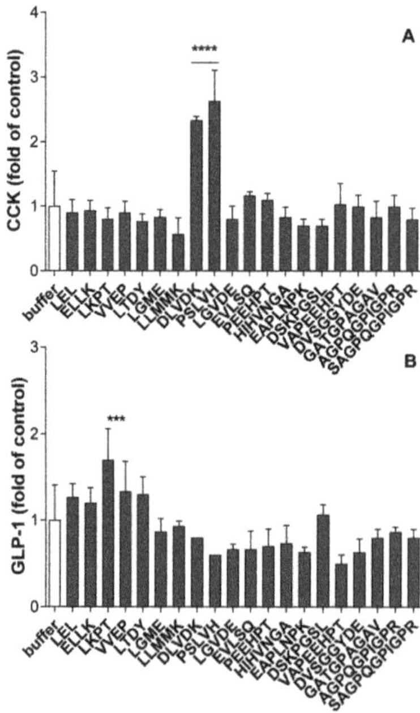

Figure 7. Synthetic peptide effects on intestinal hormones release in STC-1 cells. The amounts of intestinal hormones released in the supernatants, after 2 h of contact with the peptides (1 mM), were determined by radioimmunoassay for CCK (**A**) and active GLP-1 (**B**). Values are means of three repeated measurements and are expressed in fold of control (buffer) ± SD. Means were compared to control mean using one-way ANOVA following by a Dunnett post-hoc test, **** $p < 0.0001$; *** $p < 0.001$.

2.5. Identification of Peptides in the Basolateral Side of the Intestinal Barrier Able to Inhibit In Vitro and In Situ the DPP-IV Activity

After 2 h of contact of the FBPH intestinal digest with the Caco-2 cell monolayer in vitro IB model (apical side), 17 peptide sequences were identified by RP-UPLC-MS/MS in the basolateral side. Among these peptides, based on their presence in the most intense peaks of the UV chromatogram monitored at a wavelength of 214 nm, 13 were chemically synthesized, and their DPP-IV inhibitory activity assayed in vitro and in situ (Table 2). Results showed that five peptides (GPFPLLV, VAPEEHPT, VADTMEVV, DPLV and FAMD) were able to inhibit the DPP-IV activity in vitro, with IC_{50} values ranging from 263 to 775 µM. Seven peptides (GPFPLLV, MDLP, DLDL, FAMD, VADTMEVV, CSSGGY and VAPEEHPT) were able to inhibit the in situ DPP-IV activity with IC_{50} values ranging from 456 to 2268 µM. Four peptides (GPFPLLV, VAPEEHPT, VADTMEVV and FAMD) were able to both in vitro and in situ inhibit the DPP-IV activity.

Table 2. In vitro and in situ DPP-IV inhibitory activity of the 13selected-chemically synthesized peptides following their identification by database confrontation or de novo sequencing (ID mode). Peptides were listed according to their RP-UPLC-MS/MS characteristics (retention time (RT) and mass to charge ratio (m/z) and their identification score displayed (i) by the average local confidence (ALC) score for the de novo sequencing, (ii) by the $-10^{\log P}$ score for database confrontation with the mass error (in ppm) for both identification modes. Values of the in vitro and in situ DPP-IV inhibitory activity (IC_{50}) were determined by linear regression correlating the DPP-IV activity inhibition percentage and the Ln of the peptide concentration. nd: when the identification of the parent protein was not possible and when the IC_{50} value was above 2500 µM or undeterminable.

ID Mode	Sequence	RT	m/z	ALC	$10^{-\log P}$	In Vitro IC_{50} (µM)	In Situ IC_{50} (µM)	Identification
Database	VAPEEHPT	15.47	440.21	28.54	−3.9	409	2268	Alpha-actin
De novo	DLDL	26.88	475.24	98	−6.4	nd	763	nd
De novo	PDLV	20.77	443.24	89	−12.7	nd	nd	nd
De novo	MDLP	26.87	475.24	87	31	nd	605	nd
De novo	VDAGAP	16.52	529.26	84	0.7	nd	nd	nd
De novo	EDYT	27.16	264.10	84	−13.5	nd	nd	nd
De novo	VADTMEVV	16.06	440.21	82	4.3	603	1130	nd
De novo	DPLV	22.33	443.25	81	−8.9	698	nd	nd
De novo	EDTY	27.63	264.10	81	−14.2	nd	nd	nd
De novo	CSSGK	27.77	481.21	80	6.2	nd	nd	nd
De novo	FAMD	13.51	483.19	80	−1.4	775	862	nd
De novo	CSSGGY	29.88	573.20	80	6.4	nd	2160	nd
De novo	GPFPLLV	42.46	742.45	80	2.0	263	456	nf

3. Discussion

The first goal of this work was to study and to compare the effects of the dog gastrointestinal digestion of a tilapia byproduct protein hydrolysate and its raw material on in vitro cellular markers related to food intake and glucose homeostasis. Consequently, we first developed a static in vitro simulated dog gastrointestinal digestion based on the consensual INFOGEST protocol and on a previously one developed to study protein digestion and, according to previous works performed to investigate drug behavior [25,26] and nutrient digestibility [27] in dogs. As expected, the digestive enzymes (pepsin followed by pancreatin) exerted a more significant impact on the FBP peptide profiles than on those of FBPH, because of industrial enzymes previously digested the raw material. Although the peptide profiles and the apparent MW distribution of FBPH and FBP were quite similar at the end of the SGID, results obtained on intestinal bioactivities highlighted the added benefit of the raw material pre-hydrolysis. Indeed, the FPBH intestinal digest led to a better stimulation of active GLP-1 secretion (44.9 against 29.4 FOC) and a better inhibition of the in situ Caco-2 DPP-IV activity (5.5-fold lower IC_{50} value). Previous results obtained after the SGID of cuttlefish viscera byproduct hydrolysates and their raw material had already showed the added value of the pre-hydrolysis on the recovered intestinal digest DPP-IV inhibitory activity and GLP-1 secretion stimulation [17]. However, this was not the case for CCK secretion stimulation for which the FBPH intestinal digest was slightly less potent than the FBP one (6.2 against 7.9 FOC). The results also highlighted the crucial role of pancreatic enzymes in the apparition of protein-derived peptide bioactivities related to food intake and glucose metabolism regulation. This corroborates results obtained in previous works dealing with the SGID digestion of bovine hemoglobin and sepia byproducts on CCK and GLP-1 secretions in STC-1 cells and DPP-IV activity inhibition [17,28]. In the same way, previous works showed that intestinal digests of casein or bovine hemoglobin induced a higher stimulation of GLP-1 secretion than hydrolysate before SGID [29]. In contrast, hydrolysates may also lose their bioactivities during the gastrointestinal digestion as evidenced for a salmon skin gelatin hydrolysate which lost its GLP-1 stimulatory activity and had a significantly lower DPP-IV inhibitory activity after the SGID [30].

The FBPH intestinal digest exerted a DPP-IV inhibitory potential characterized by an IC_{50} value equal to 1.52 mg·mL^{-1} when obtained by in vitro biochemical test. This is in line with numerous studies showing IC_{50} values for marine byproduct hydrolysates ranging from 1 to 5 mg·mL^{-1} [17,19,22,31,32] and even less than 1 mg·mL^{-1} as for *Gadus chalcogrammus* gelatin [33] and *Salmo salar* hydrolysates [30]. Here, we used an in situ DPP-IV activity test using live Caco-2 cells, which mimics the intestinal environment and, in particular, the enzymatic action of peptidases produced by the epithelial cells of the intestinal brush border [34]. The IC_{50} value obtained for the FBPH intestinal digest was 0.67 mg·mL^{-1}. This value is obviously not comparable with those obtained with the in vitro classical test. Nevertheless, this DPP-IV inhibitory activity appears very promising when compared with the IC_{50} (1.57 mg·mL^{-1}) obtained for an intestinal cuttlefish byproduct hydrolysate digest in a previous work with the same Caco-2 in situ test [34].

The in vitro results obtained here can be extrapolated to those previously obtained in vivo with other fish protein hydrolysates on food intake and glycemic managements in healthy mice [20] and rats [16], in high-fat-diet-induced obese mice [35] or diabetic and obese rats and as in several clinical studies [18,21,33,36–38].

To identify, from the FBPH intestinal digest, active peptides able to stimulate intestinal hormones secretion by EECs, a methodology built on two different successive separation techniques were adopted Caron, J. et al., 2016 [23,25]. Using a first SEC-purification step, the fraction F2 composed by a majority of peptides characterized by apparent MW ranging from 400 to 1000 Da was selected on the basis of its intestinal hormones release activity and submitted to RP-HPLC. Finally, the FE subfraction obtained after RP-HPLC separation exerted the best stimulating release effect for both intestinal hormones, unambiguously.

Among the 1739 peptide sequences identified by the mass bioinformatics data processing, 20 of them were selected (based on their presence in the most intense peaks of the UV chromatogram (λ = 214 nm) and their ion intensity and fragmentation quality), chemically synthesized and assayed for their capacity to stimulate CCK and GLP-1 release. The results allowed to identify two new peptides, PSLVH and DLVDK, able to enhance CCK release by EECs, and one tetrapeptide, FAMD, able to stimulate GLP-1 release. Today, few peptide sequences are reported in the literature to stimulate CCK and GLP-1 secretion by EECs, and the relationship existing between the CCK and GLP-1 food-derived releasing peptide bioactivity and their structure and amino acid sequence is not well established [9,38]. Nevertheless, different signaling pathways, involving G protein-coupled receptors (GPCRs) like GPR93, GPRC6A and the calcium-sensing receptor (CaSR) but also the cotransporter PepT-1, have been evidenced in the food protein-derived peptide intestinal sensing leading to CCK and GLP-1 secretion [39–41]. CCK releasing food-derived peptides were identified from soybean β-conglycinin, bovine hemoglobin, lactoglobulin, bovine whey, casein, and egg white protein [42–48]. To our knowledge, it is the first time from fish source. The motif and the structure of the peptides appear crucial in the intestinal sensing leading to CCK secretion. CaSR was described to sense W and F aromatic amino acids [49,50] and the presence of aromatic residues in the peptide sequence seems to favor the bioactivity. In previous works, we evidenced two CCK-releasing fractions of a bovine hemoglobin SGID intestinal digest. They were able to highly stimulate CCK secretion and composed of more than 50% of peptides containing at least one aromatic amino acid residue in their sequence. The four hemorphins (LLVVYPWT, LVVYPWT, VVYPWT and VVYPWTQRF), released during bovine hemoglobin digestion, were synthesized and proved as CCK and GLP-1 secretion stimulating peptides [42,44]. However, in the present work, the two identified CCK-releasing stimulating peptides, PSLVH and DLVDK, do not possess aromatic residues, whereas DVSGGYDE did not stimulate CCK secretion. These two active peptides are both composed of five amino acid residues and some of them contain an aliphatic chain. These findings are in accordance with a precedent work which hypothesized that five amino acid residues were the minimal size and that the presence of aliphatic chain could be crucial in the CCK secretion in STC-1 cells [47]. In accordance, in a recent work, two peptides able to stimulate CCK secretion in STC-1 cells, VLLPDEVSGL and VLLPD, were identified from an egg white SGID intestinal digest. They both did not contain aromatic amino acid residues but are characterized by a high rate of aliphatic ones [48].

Like for CCK, only few food-derived peptides, able to stimulate the secretion of GLP-1, were identified. We previously identified four sequences (KAAVT, TKAVEH, ANVST and YGAE) from a bovine hemoglobin intestinal digest and proposed that the presence of basic amino acid residue (L-lysine) in the N-terminal side of the peptide, as well as the presence of a T residue in the C- or N-terminal, are common points that could be implied in the peptide sensing that led to GLP-1 secretion [42]. LKPT evidenced in the present work also possesses a lysine amino acid residue in N-terminal position as it was also found in the minimal sequence from α-actinin-2 (KPYIL) able to stimulate the GLP-1 secretion in murine GLUTag cells. However, the K residue position in the sequence does not seem crucial for the bioactivity as ASDKPYIL is also active [51]. The RVASMASEKM peptide, recently identified from egg white protein digest as GLP-1 secretagogue, also possesses a K residue but in C-terminal position [48]. Nevertheless, results obtained here showed that ELLK and EAPLNPK did not lead to GLP-1 secretion, and other works identified peptides able to stimulate GLP-1 secretion without having K or T residues in their sequences, such as GGGG, AAAA, GWGG [52], GPVRGPFPIIV [53], LGG and GF [54] and, PFL [48].

Taken together, these findings confirm the presence of multiple pathways involved in the intestinal peptide sensing leading to CCK and GLP-1 secretion by EECs. It will be necessary to identify which one is used by each peptide to elucidate the relationships between the physicochemical properties, the structure, and the sequence of the peptide and its secretagogue activity.

Among the 13 synthesized peptides identified in the basolateral side of the intestinal barrier model, 5 exerted an in vitro DPP-IV inhibitory activity, with IC_{50} values ranging from 263 to 775 µM (Table 2). These peptides are promising compared with food-derived DPP-IV inhibitory peptides identified between 2016 and 2018 as recently reviewed by Liu et al. [55]. Indeed, when we perform from the Liu et al. list, the analysis of 74 peptides, identified and characterized by IC_{50} values ranging from 43 to 2000 µM, the IC_{50} mean and median values were 596 and 226 µM, respectively. The large majority of the studies which identified DPP-IV inhibitory dietary protein-derived peptides has used in vitro controlled method to calculate IC_{50} values and did not assay the ability of the peptides to cross the intestinal barrier. In the recent work of Harnedy et al., the authors evaluated the DPP-IV inhibitory activity potential of peptides identified from two RP-HPLC fractions of a boarfish hydrolysate submitted to SGID. Several peptides were then synthesized and assayed for their in vitro inhibitory DPP-IV activity. The most promising peptides ($IC_{50} < 200$ µM) were further assayed for their ability to inhibit in situ the human DPP-IV activity using culture Caco-2 cells in order to better mimic intestinal physiological conditions. They indeed identified 18 peptides with in situ IC_{50} values ranging from 44 to 307 µM [32]. Despite these very interesting findings, the ability of the identified peptides to cross the IB still needs to be studied. Indeed, several studies showed that many DPP-IV inhibitory peptides identified in the intestinal tract cannot cross the IB without being cleaved and losing their bioactivities. Other studies showed that certain DPP-IV inhibitory peptides were able to cross the IB in vitro. Domenger et al. showed that among five DPP-IV inhibitory peptides identified in a bovine hemoglobin intestinal digest, only three of them were recovered intact after the passage through a Caco-2 cells monolayer [44,56]. Lacroix et al., also evidenced the susceptibility of certain milk protein-derived DPP-IV inhibitory peptides to be cleaved by brush barrier peptidases [57]. Indeed, the differentiated Caco-2 cells express mainly two peptidases, DPP-IV and transmembrane protease serine 4 (TMPRSS4) which have been evidenced to hydrolyze peptides during the passage through the simulated IB [58].

In the present study, we adopted the strategy to first incubate the whole digested hydrolysate at the apical side of the IB model for 2 h in order to further identify the peptides in the basolateral compartment. Adopting this strategy also permitted to mimic the interaction of the whole digest with the IB which may modify its permeability. There is some evidence that food-derived peptides could alter intestinal barrier permeability via their actions on tight junction proteins [59,60]. Indeed, we evidenced in a previous work, four hemorphins harboring DPP-IV inhibitory activity able to significantly decrease mRNA expression of the claudin 4, a protein present in tight junctions and involved in paracellular permeability [56]. Finally, the eight new DPP-IV inhibitory peptides which were evidenced in this study might be able in vivo to reach the plasmatic compartment in a sufficient concentration and to inhibit the DPP-IV circulating form, enhancing the half-life of the GLP-1 and therefore its incretin and satiating actions. Moreover, it is crucial to keep in mind that a substantial number of potentially bioactive peptides, in particular small ones, are unidentifiable due to the current peptidomics advance [61]. Further in vivo studies are needed to evidence the glucose and/or food intake regulatory effect of this FBPH, nevertheless, the present tilapia byproduct hydrolysate appears to be very promising as functional ingredient preventing or managing overweight and glucose tolerance.

4. Materials and Methods

4.1. Materials and Chemicals

Porcine pepsin (EC number 3.4.23.1, from porcine gastric mucosa, >250 U mg^{-1}), pancreatin from porcine pancreas (4× USP specifications, European commission number: 232-468-9), diprotin A, Gly-Pro-7-amido-4-methylcoumarin hydrobromide (H-Gly-Pro-AMC, HBr) were purchased from Sigma-Aldrich (Villefranche-sur-Saône, France). Dulbecco's modified Eagle's cell culture Medium (DMEM), trypsin, L-glutamine, fetal bovine serum, antibiotics (penicillin and streptomycin) were purchased from Dutscher (Issy-les-Moulineaux, France). Fish byproduct protein hydrolysate (FBPH) was obtained by the hydrolysis of tilapia (*Oreochromis niloticus*) byproducts (FBP, bones and viscera), using commercially available food grade enzymes. FBPH and FBP were provided by Diana Pet Food (Elven, France). Synthetic peptides were purchased from GeneCust (Boynes, France).

4.2. In Vitro Simulated Canine Gastrointestinal Digestion of FBP and FBPH

The simulated gastrointestinal digestion (SGID) was adapted from the static in vitro consensual protocol coming from the INFOGEST cost action (http://www.cost-infogest.eu), as well as from Caron et al. in order to mimic the dog gastrointestinal digestion [24,28]. Briefly, the three first steps of the digestive tract (oral, gastric and intestinal) were simulated using a static mono compartmental process and under constant magnetic stirring in a reactor at 39 °C. Two grams of FBPH or FBP were solubilized in 16 mL of salivary fluid at pH 7.0 without salivary enzyme. A 4 mL aliquot (oral aliquot) was withdrawing after 2 min. Twenty-four mL of gastric fluids were then added before the addition of porcine pepsin in a 1:40 (*w/v*) E/S ratio (enzymatic activity > 2000 U mg^{-1} of dry weight). Gastric digestion was performed over 2 h, pH being monitored and maintained at pH 2.0 with NaOH (5 M) and HCl (5 M) solutions. Hydrolysate aliquots (gastric aliquots) were withdrawing after 2 h and directly heated at 95 °C during 10 min. Thirty-six mL of intestinal fluid and 4 mL of 1 M NaHCO$_3$ solution were added to reach the pH to 6.8. Pancreatin was added in a 1:50 (*w/v*) ratio E/S (enzymatic activity 100 U mg^{-1} of dry weight) and intestinal digestion was carried out over 4 h. Aliquots (intestinal aliquots) were withdrawn and heated as above. All aliquots were then centrifuged at $13,000 \times g$ for 10 min and supernatants were collected to be stored at $-20\,^\circ$C for further analysis.

4.3. Size Exclusion Chromatography by Fast Protein Liquid Chromatography (SEC-FPLC)

The peptide apparent molecular weight (MW) distributions of oral, gastric and intestinal aliquots were obtained by SEC using a Superdex Peptide 10/300 GL column (GE Healthcare, Uppsala, Sweden) on an AKTA Purifier system (GE Healthcare). SEC was carried out in isocratic conditions with an elution solution of 30% acetonitrile, 69.9% ultrapure water and 0.1% TFA solvent at a flow rate of 0.5 mL.min^{-1}. Oral, gastric and intestinal aliquots were first diluted in ultrapure water (18.5 g L^{-1}, w/v) and subjected to a magnetic stirring for 15 min. The diluted samples were then centrifuged at $15,000 \times g$ for 15 min and the supernatants were filtered through a 0.22 µm membrane filter before injection. The absorbance was monitored at 214 nm for 70 min. The column was calibrated with the following standard peptides: cytochrome C (12,327 Da), aprotinin (6511 Da), insulin beta-chain (3496 Da), neurotensin (1673 Da), substance P (1348 Da), substance P fragment 1–7 (900 Da) and leupeptin (463 Da).

4.4. Cell Culture Conditions

The Caco-2 cell line was purchased from Sigma-Aldrich (Villefranche-sur-Saône, France) and the STC-1 cell line was a grateful gift received from Corinne Grangette (Univ. Lille, CNRS, Inserm, CHU Lille, Institut Pasteur de Lille, U1019-UMR 8204-CIIL, France). Cells were grown in flask of 75 cm^2 at 37 $^\circ$C, 5% CO$_2$ atmosphere in DMEM supplemented with 4.5 g L^{-1} of glucose, 10% of fetal bovine serum, 100 U mL^{-1} of penicillin, 100 µg mL^{-1} of streptomycin and 2 mM of L-glutamine. Caco-2 and STC-1 cells were weekly and twice a week subcultured, respectively. All cells used in this study were between the 40 and the 50 passages for Caco-2 cells and between the 10 and 30 passages for STC-1 cells.

4.5. CCK and GLP-1 Secretion Study

When 80–90% confluence was reached, STC-1 cells were trypsinized and seeded at a density of 40,000 cells/well in 24-wells culture plates (ThermoFisher Scientific, Saint Aubin, France) allowing to reach 60–80% confluence. Cell culture medium was removed from each well and cells were washed with phosphate saline buffer (PBS, 10 mM, pH 7.4). 250 µL of digests (2 to 10 mg mL^{-1}) or synthetic peptides (1 mM) diluted in Hepes buffer (4.5 mM KCl, 1.2 mM CaCl$_2$, 140 mM NaCl and 20 mM Hepes, pH 7.4) were then added. Hepes buffer was used as a negative control. After 2 h of incubation at 37 $^\circ$C, 5% CO$_2$ atmosphere, supernatants were collected on ice, centrifuged ($1500 \times g$ for 5 min) and stored at $-20\,^\circ$C for further CCK and GLP-1 concentration measurements using GASK-PR (Cisbio, Codolet, France) and GLP-1 active (Merck, Molsheim, France) RIA kits, respectively.

4.6. DPP-IV Activity Assay

In situ method using confluent Caco-2 cells described by Caron et al. was slightly modified and used to study DPP-IV activity [34]. A 1 mM (Gly-Pro-AMC) substrate solution, the digests and the synthetic peptides dilutions were prepared in phosphate saline buffer pH 7.4 (PBS). Briefly, after 7 days of growth, Caco-2 cells were trypsinized and seeded at a density of 8000 cells/well in 96-well optical black plates (Nunc, ThermoFisher Scientific, Rochester, NY, USA). After 7 days, culture media were removed from wells and the cells were washed with 100 µL of PBS buffer (pH 7.4). Then, 100 µL of PBS was added to the wells followed by 25 µL of digests diluted in PBS at increasing concentrations (3.47, 6.95 and 13.89 mg mL^{-1}) or 25 µL of synthetic peptides diluted in PBS at increasing concentrations (between 0.2; 0.6; 1 and 1.5 mM) or PBS buffer (control wells). After 5 min of incubation at 37 $^\circ$C, 50 µL of (Gly-Pro-AMC) substrate solution were added to each well. Fluorescence was recorded every 2 min for 1 h at 37 $^\circ$C using a Xenius XC spectrofluorometer (Safas Monaco, Monaco). Excitation wavelength was set to 260 nm while the emission wavelength was of 480 nm. The percentage of the DPP-IV activity inhibition was defined as the percentage of DPP-IV activity inhibited by a given concentration of digest or diprotin A (commercial DPP-IV peptide inhibitor) as positive control compared with control buffer response. The concentration of digests or synthetic peptide solutions required to obtain 50% inhibition of the DPP-IV activity (IC_{50}) was determined by plotting the percentage of DPP-IV activity inhibition as a function of digest or peptide final concentration natural logarithm. IC_{50} was expressed in mg mL^{-1} or in mM.

4.7. Fractionation of the FBPH Intestinal Digest
4.7.1. SEC-FPLC Fractionation

The intestinal peptide population of FBPH duodenal SGID was separated on an Akta Purifer device using a preparative HiLoad 16/600 Superdex 30 prepgrade column (GE Healthcare). A volume of 2 mL at 18.5 mg mL^{-1} dry matter of the intestinal digest was injected in the column and eluted in isocratic conditions with an eluent composed of 30% acetonitrile, 69.9% of ultrapure water and 0.1% of TFA at 1 mL.min^{-1} for 2 h. The collected fractions were then dried by centrifuge evaporation (MiVac Quattro Concentrator, Biopharma Process Systems, Winchester, UK) and the obtained pellets were re-solubilized in 1 mL of ultrapure water and stocked at $-20\ ^{\circ}$C.

4.7.2. HPLC Fractionation

The FPLC fractions displaying the strongest bioactivities was fractionated with a semi-preparative C18 Gemini column (150 × 10 mm, particles size 5 μm, 110 Å, Phenomenex, Le Pecq, France) on a 4250 Puriflash system (Interchim, Montluçon, France). The peptide elution was performed at a flow rate of 5 mL.min^{-1} with two solvents: eluent A was composed of 99.9% of ultrapure water and 0.1% TFA and eluent B was composed of 99.9% of acetonitrile and 0.1% TFA. The following hydrophobic gradient was used: an isocratic step at 98% of eluent A for 20 min followed by a linear gradient from 2% to 15% of eluent B in 35 min, then a linear gradient from 15% to 90% of eluent B in 10 min and finally the column was washed with 90% of eluent B for 5 min and equilibrated again with 98% of eluent A for 10 min. The collected subfractions were then dried by centrifuge evaporation (MiVac Quattro Concentrator, Biopharma Process Systems).

4.8. Peptide Sequences Identification in HPLC Fractions
4.8.1. RP-HPLC-MS/MS Analysis of HPLC Fractions

Selected dried HPLC subfractions were re-solubilized in 50 μL of ultrapure water containing 0.1% of formic acid (FA), vortexed, submerged in ultrasonic bath three times and finally centrifuged 5 min at 12,000× *g*. The peptides of these fractions (10 μL injection volume) were then chromatographed by reverse phase-ultra high-performance liquid chromatography (RP-UPLC) using an ACQUITY biocompatible chromatography system (Waters, Manchester, UK) equipped with an analytical C18 Uptisphere column (250 × 3 mm, particles size 5 μm, 300 Å, Interchim). The peptides elution was performed at 30 $^{\circ}$C with a flow rate of 0.6 mL.min^{-1} using two solvents: eluent A was composed of 99.9% of ultrapure water and 0.1% FA and eluent B of 99.9% of acetonitrile and 0.1% FA. Apolar elution gradient used was: 100% of eluent A for 2 min followed by a linear gradient from 0 to 15% of eluent B in 45 min, then a linear gradient from 15% to 35% of eluent B in 20 min and from 35% to 90% of eluent B in 15 min. The column was finally washed with 90% of eluent B for 10 min and equilibrated with 100% of eluent A for 7 min.

The chromatographed peptides were then ionized into the electrospray ionization source of the qTOF Synapt G2-Si™ (Waters). MS analysis was performed in sensitivity, positive ion and data dependent analysis (DDA) modes. The source temperature was set at 150 $^{\circ}$C, the capillary and cone voltages were set at 3000 and 60 V, respectively. MS and MS/MS measurements data were performed in a mass/charge range fixed between 100 to 2000 *m/z* with a scan time of 0.2 s. A maximum of 15 precursor ions with an intensity threshold of 10,000 were selected for the fragmentation by collision induced dissociation

(CID) with specified voltages ranging from 8 to 9 V and from 40 to 90 V for the lower molecular mass ions and for those with a higher molecular mass, respectively. The leucin enkephalin ($[M + H]^+$ of 556.632) was injected in the system every 2 min for 0.5 s to follow and to correct the measure error during all the time of analyze.

4.8.2. Mass Spectrometry Data Processing

Mass spectrometry data processing and the protein database search were performed via Peaks Studio version 8.5 software (Bioinformatics Solutions, Waterloo, ON, Canada) using UniProt database restricted to the complete proteome of the Cichlidae family (updated the 2018/08/28, 44,684 entries). Tolerance threshold of precursor ion masses and fragments were defined at 35 ppm and 0.2 Da, respectively. The in-database identification search was performed with consideration of oxidized methionine but without notifying the choice of enzyme. Peptide sequences identified by the Peaks Studio 8.5 were filtered with a fault discovery rate (FDR) strictly lower than 1% while peptide sequences identified by de novo processing were filtered according to an average local confidence score (ALC score) up to 80%.

4.9. Intestinal Barrier Passage and Peptide Identification

4.9.1. Transport Study

To obtain an intestinal barrier (IB) model, Caco-2 cells were cultivated on insert (D = 4.2 cm; pore size = 3 μm, ref: 353092, Dutscher) in 6-wells plate where Caco-2 cells were seeded at a density of 84,000 cells by insert in 2 mL of DMEM. A volume of 2.5 mL of DMEM medium was added in each well of the plate. Cells were incubated at 37 °C for 15 days and the medium on apical and basolateral sides were changed every 2 days.

In the day of experimentation, a transport medium Hepes-Hanks salt solution (HBSS) was extra temporary prepared and filtered on PVDF filter 0.22 μm. Samples were diluted at 4 g L^{-1} with the transport medium. The apical and basolateral sides of each well were washed with 500 μL and 1 mL of transport medium (heated at 37 °C), respectively. Then, 1 mL of transport medium at 37 °C was added in apical side and 2.5 mL in basolateral side. Plate was incubated at 37 °C, 5% of CO_2 for 30 min, and the supernatant was discarded and replaced with 1 mL of pre-heated samples or pre-heated transport medium (for the control). Kinetic studies were performed by sampling 100 μL from the apical and basolateral sides at 15 min, 250 μL in apical side and 1 mL in basolateral side at 60 min and the rest of the supernatant in apical (650 μL) and in basolateral side (1.4 mL) at 120 min of incubation at 37 °C, 5% CO_2. The peptides were identified after 120 min incubation time.

4.9.2. Peptide Sequences Identification in Apical and Basolateral Supernatant by Mass Spectrometry

Apical and basolateral supernatants at 120 min were prepared and analyzed by mass spectrometry with the same protocol described above for HPLC fractions with minor changes. The UPLC column used was a C18-AQ (150 × 3 mm, particles size: 2.6 μm, 83 Å, Interchim) and the peptide chromatography was performed at a flow rate of 0.5 mL·min^{-1} and 30 °C. The apolar elution gradient used was as follow: 5 min at 99% of eluent A/1% eluent B, then a linear gradient from 1% to 30% of eluent B in 40 min, followed by a linear gradient from 30% to 70% of eluent B in 8 min, and finally after 2 min at 95% of eluent B, the column was equilibrated with 99% of eluent A/1% eluent B for 3 min. The ionization mode, and the MS and MS/MS measures were performed exactly as described previously.

4.10. Statistical Analysis

Data presented are means ± SD. To compare GI hormone secretion levels induced by the digests, a one-way ANOVA using general linear model and pairwise comparisons with Tukey's or Dunnett's tests were performed using Graph Prism (GraphPad Software, San Diego, CA, USA). Values were considered as significantly different for a p-value < 0.05.

5. Conclusions

A dog in vitro static simulated gastrointestinal digestion model that permitted us to evaluate in vitro the potential effects of a tilapia byproduct hydrolysate on the regulation of food intake and glucose metabolism was developed. Promising effects on intestinal hormones secretion and dipeptidyl peptidase IV (DPP-IV) inhibitory activity were thus evidenced and the added-value of the pre-hydrolysis was highlighted. New bioactive peptides able to stimulate CCK (DLVDK and PSLVH) and GLP-1 (LKPT) secretion and to inhibit the DPP-IV activity after a transport study through an intestinal barrier (VAPEEHPT, DLDL, MDLP, VADTMEVV, DPLV, FAMD, CSSGGY and GPFPLLV) were identified. This tilapia byproduct hydrolysate appears to be promising to manage overweight.

Author Contributions: Conceptualization, B.C., R.R., A.L.; methodology, S.T., B.C., R.R., B.D. and C.F.; formal analysis, S.T., B.D., B.C., R.R. and C.F.; investigation, S.T., C.P., B.C., and B.D.; data curation, S.T., B.D., B.C., and C.F.; writing original draft preparation, S.T and B.C.; writing review and editing, B.C., B.D., A.L., C.F., and R.R.; supervision, R.R., B.C., A.L., I.G. All authors have read and agreed to the published version of the manuscript.

Funding: This work has been carried out in the framework of Alibiotech research program, which is financed by European Union, French State, and the French Region of Hauts-de-France. The HPLC-MS/MS experiments were performed on the REALCAT platform funded by a French governmental subsidy managed by the French National Research Agency (ANR) within the frame of the "Future Investments' program (ANR-11-EQPX-0037)".

Data Availability Statement: The data presented in this study are available on request from the corresponding author. The data are not publicly available due to this study is an industrial work and some results are confidential.

Conflicts of Interest: The raw material and the hydrolysate have been provided by Diana Pet Food.

Sample Availability: Samples of the compounds are not available from the authors.

References

1. United Nations; Department of Economic and Social Affairs; Population Division. *World Population Prospects Highlights*; UN: New York, NY, USA, 2019; ISBN 978-92-1-148316-1.
2. Henchion, M.; Hayes, M.; Mullen, A.M.; Fenelon, M.; Tiwari, B. Future Protein Supply and Demand: Strategies and Factors Influencing a Sustainable Equilibrium. *Foods* **2017**, *6*, 53. [CrossRef] [PubMed]
3. FAO. The State of World Fisheries and Aquaculture 2020: Sustainability in Action. In *The State of World Fisheries and Aquaculture (SOFIA)*; FAO: Rome, Italy, 2020; ISBN 978-92-5-132692-3.
4. Saeedi, P.; Petersohn, I.; Salpea, P.; Malanda, B.; Karuranga, S.; Unwin, N.; Colagiuri, S.; Guariguata, L.; Motala, A.A.; Ogurtsova, K.; et al. Global and Regional Diabetes Prevalence Estimates for 2019 and Projections for 2030 and 2045: Results from the International Diabetes Federation Diabetes Atlas, 9th Edition. *Diabetes Res. Clin. Pract.* **2019**, *157*, 107843. [CrossRef] [PubMed]
5. Chandler, M.; Cunningham, S.; Lund, E.M.; Khanna, C.; Naramore, R.; Patel, A.; Day, M.J. Obesity and Associated Comorbidities in People and Companion Animals: A One Health Perspective. *J. Comp. Pathol.* **2017**, *156*, 296–309. [CrossRef] [PubMed]
6. Öhlund, M.; Palmgren, M.; Holst, B.S. Overweight in Adult Cats: A Cross-Sectional Study. *Acta Vet. Scand.* **2018**, *60*, 5. [CrossRef] [PubMed]
7. Janssen, S.; Depoortere, I. Nutrient Sensing in the Gut: New Roads to Therapeutics? *Trends Endocrinol. Metab.* **2013**, *24*, 92–100. [CrossRef] [PubMed]
8. Rønnestad, I.; Akiba, Y.; Kaji, I.; Kaunitz, J.D. Duodenal Luminal Nutrient Sensing. *Curr. Opin. Pharmacol.* **2014**, *19*, 67–75. [CrossRef]
9. Caron, J.; Domenger, D.; Dhulster, P.; Ravallec, R.; Cudennec, B. Protein Digestion-Derived Peptides and the Peripheral Regulation of Food Intake. *Front. Endocrinol.* **2017**, *8*. [CrossRef]
10. Nauck, M.A.; Meier, J.J. Incretin Hormones: Their Role in Health and Disease. *Diabetes Obes. Metab.* **2018**, *20*, 5–21. [CrossRef]
11. Engel, M.; Hoffmann, T.; Manhart, S.; Heiser, U.; Chambre, S.; Huber, R.; Demuth, H.-U.; Bode, W. Rigidity and Flexibility of Dipeptidyl Peptidase IV: Crystal Structures of and Docking Experiments with DPIV. *J. Mol. Biol.* **2006**, *355*, 768–783. [CrossRef]
12. Godinho, R.; Mega, C.; Teixeira-de-Lemos, E.; Carvalho, E.; Teixeira, F.; Fernandes, R.; Reis, F. The Place of Dipeptidyl Peptidase-4 Inhibitors in Type 2 Diabetes Therapeutics: A "Me Too" or "the Special One" Antidiabetic Class? *J. Diabetes Res.* **2015**, *2015*, 28. [CrossRef]
13. Nongonierma, A.B.; FitzGerald, R.J. Prospects for the Management of Type 2 Diabetes Using Food Protein-Derived Peptides with Dipeptidyl Peptidase IV (DPP-IV) Inhibitory Activity. *Curr. Opin. Food Sci.* **2016**, *8*, 19–24. [CrossRef]

14. Dale, H.F.; Madsen, L.; Lied, G.A. Fish–Derived Proteins and Their Potential to Improve Human Health. *Nutr. Rev.* **2019**, *77*, 572–583. [CrossRef] [PubMed]

15. Cudennec, B.; Ravallec-Plé, R.; Courois, E.; Fouchereau-Peron, M. Peptides from Fish and Crustacean By-Products Hydrolysates Stimulate Cholecystokinin Release in STC-1 Cells. *Food Chem.* **2008**, *111*, 970–975. [CrossRef]

16. Cudennec, B.; Fouchereau-Peron, M.; Ferry, F.; Duclos, E.; Ravallec, R. In Vitro and in Vivo Evidence for a Satiating Effect of Fish Protein Hydrolysate Obtained from Blue Whiting (Micromesistius Poutassou) Muscle. *J. Funct. Foods* **2012**, *4*, 271–277. [CrossRef]

17. Cudennec, B.; Balti, R.; Ravallec, R.; Caron, J.; Bougatef, A.; Dhulster, P.; Nedjar, N. In Vitro Evidence for Gut Hormone Stimulation Release and Dipeptidyl-Peptidase IV Inhibitory Activity of Protein Hydrolysate Obtained from Cuttlefish (Sepia Officinalis) Viscera. *Food Res. Int.* **2015**, *78*, 238–245. [CrossRef]

18. Nobile, V.; Duclos, E.; Michelotti, A.; Bizzaro, G.; Negro, M.; Soisson, F. Supplementation with a Fish Protein Hydrolysate (Micromesistius Poutassou): Effects on Body Weight, Body Composition, and CCK/GLP-1 Secretion. *Food Nutr. Res.* **2016**, *60*, 29857. [CrossRef]

19. Harnedy, P.A.; Parthsarathy, V.; Mclaughlin, C.M.; Kee, M.B.O.; Allsopp, P.J.; Mcsorley, E.M.; Harte, F.P.M.O.; Fitzgerald, R.J. Blue Whiting (Micromesistius Poutassou) Muscle Protein Hydrolysate with in Vitro and in Vivo Antidiabetic Properties. *J. Funct. Foods* **2018**, *40*, 137–145. [CrossRef]

20. Parthsarathy, V.; McLaughlin, C.M.; Harnedy, P.A.; Allsopp, P.J.; Crowe, W.; McSorley, E.M.; FitzGerald, R.J.; O'Harte, F.P.M. Boarfish (Capros Aper) Protein Hydrolysate Has Potent Insulinotropic and GLP-1 Secretory Activity in Vitro and Acute Glucose Lowering Effects in Mice. *Int. J. Food Sci. Technol.* **2019**, *54*, 271–281. [CrossRef]

21. Hsieh, C.H.; Wang, T.Y.; Hung, C.C.; Chen, M.C.; Hsu, K.C. Improvement of Glycemic Control in Streptozotocin-Induced Diabetic Rats by Atlantic Salmon Skin Gelatin Hydrolysate as the Dipeptidyl-Peptidase IV Inhibitor. *Food Funct.* **2015**. [CrossRef]

22. Huang, S.L.; Jao, C.L.; Ho, K.P.; Hsu, K.C. Dipeptidyl-Peptidase IV Inhibitory Activity of Peptides Derived from Tuna Cooking Juice Hydrolysates. *Peptides* **2012**, *35*, 114–121. [CrossRef]

23. Li-chan, E.C.Y.; Hunag, S.; Jao, C.; Ho, K.; Hsu, K. Peptides Derived from Atlantic Salmon Skin Gelatin as Dipeptidyl-Peptidase IV Inhibitors. *J. Agric. Food Chem.* **2012**, *60*, 973–978. [CrossRef] [PubMed]

24. Minekus, M.; Alminger, M.; Alvito, P.; Ballance, S.; Bohn, T.; Bourlieu, C.; Carriere, F.; Boutrou, R.; Corredig, M.; Dupont, D. A Standardised Static in Vitro Digestion Method Suitable for Food–an International Consensus. *Food Funct.* **2014**, *5*, 1113–1124. [CrossRef] [PubMed]

25. Arndt, M.; Chokshi, H.; Tang, K.; Parrott, N.J.; Reppas, C.; Dressman, J.B. Dissolution Media Simulating the Proximal Canine Gastrointestinal Tract in the Fasted State. *Eur. J. Pharm. Biopharm.* **2013**, *84*, 633–641. [CrossRef] [PubMed]

26. Hatton, G.B.; Yadav, V.; Basit, A.W.; Merchant, H.A. Animal Farm: Considerations in Animal Gastrointestinal Physiology and Relevance to Drug Delivery in Humans. *J. Pharm. Sci.* **2015**, *104*, 2747–2776. [CrossRef]

27. Hervera i Abab, M. Methods for Predicting the Energy Value of Commercial Dog Foods. Ph.D. Thesis, Universitat Autònoma de Barcelona, Barcelona, Spain, 2011.

28. Caron, J.; Domenger, D.; Belguesmia, Y.; Kouach, M.; Lesage, J.; Goossens, J.-F.; Dhulster, P.; Ravallec, R.; Cudennec, B. Protein Digestion and Energy Homeostasis: How Generated Peptides May Impact Intestinal Hormones? *Food Res. Int.* **2016**, *88*, 310–318. [CrossRef]

29. Chaudhari, D.D.; Singh, R.; Mallappa, R.H.; Rokana, N.; Kaushik, J.K.; Bajaj, R.; Batish, V.K.; Grover, S. Evaluation of Casein & Whey Protein Hydrolysates as Well as Milk Fermentates from Lactobacillus Helveticus for Expression of Gut Hormones. *Indian J. Med. Res.* **2017**, *146*, 409–419. [CrossRef]

30. Harnedy, P.A.; Parthsarathy, V.; McLaughlin, C.M.; O'Keeffe, M.B.; Allsopp, P.J.; McSorley, E.M.; O'Harte, F.P.M.; FitzGerald, R.J. Atlantic Salmon (Salmo Salar) Co-Product-Derived Protein Hydrolysates: A Source of Antidiabetic Peptides. *Food Res. Int.* **2018**, *106*, 598–606. [CrossRef]

31. Wang, T.-Y.; Hsieh, C.-H.; Hung, C.-C.; Jao, C.-L.; Chen, M.-C.; Hsu, K.-C. Fish Skin Gelatin Hydrolysates as Dipeptidyl Peptidase IV Inhibitors and Glucagon-like Peptide-1 Stimulators Improve Glycaemic Control in Diabetic Rats: A Comparison between Warm- and Cold-Water Fish. *J. Funct. Foods* **2015**, *19*, 330–340. [CrossRef]

32. Harnedy-Rothwell, P.A.; McLaughlin, C.M.; O'Keeffe, M.B.; Le Gouic, A.V.; Allsopp, P.J.; McSorley, E.M.; Sharkey, S.; Whooley, J.; McGovern, B.; O'Harte, F.P.M.; et al. Identification and Characterisation of Peptides from a Boarfish (Capros Aper) Protein Hydrolysate Displaying in Vitro Dipeptidyl Peptidase-IV (DPP-IV) Inhibitory and Insulinotropic Activity. *Food Res. Int.* **2020**, *131*, 108989. [CrossRef]

33. Guo, L.; Harnedy, P.A.; Zhang, L.; Li, B.; Zhang, Z.; Hou, H.; Zhao, X.; FitzGerald, R.J. In Vitro Assessment of the Multifunctional Bioactive Potential of Alaska Pollock Skin Collagen Following Simulated Gastrointestinal Digestion. *J. Sci. Food Agric.* **2015**, *95*, 1514–1520. [CrossRef]

34. Caron, J.; Domenger, D.; Dhulster, P.; Ravallec, R.; Cudennec, B. Using Caco-2 Cells as Novel Identi Fi Cation Tool for Food-Derived DPP-IV Inhibitors. *Food Res. Int.* **2017**, *92*, 113–118. [CrossRef] [PubMed]

35. Wang, Y.; Gagnon, J.; Nair, S.; Sha, S. Herring Milt Protein Hydrolysate Improves Insulin Resistance in High-Fat-Diet-Induced Obese Male C57BL/6J Mice. *Mar. Drugs* **2019**, *17*, 456. [CrossRef] [PubMed]

36. Crowe, W.; McLaughlin, C.M.; Allsopp, P.J.; Slevin, M.M.; Harnedy, P.A.; Cassidy, Y.; Baird, J.; Devaney, M.; Fitzgerald, R.J.; O'Harte, F.P.M.; et al. The Effect of Boarfish Protein Hydrolysate on Postprandial Glycaemic Response and Satiety in Healthy Adults. *Proc. Nutr. Soc.* **2018**, *77*. [CrossRef]

37. Dale, H.F.; Jensen, C.; Hausken, T.; Lied, E.; Hatlebakk, J.G.; Brønstad, I.; Lihaug Hoff, D.A.; Lied, G.A. Effect of a Cod Protein Hydrolysate on Postprandial Glucose Metabolism in Healthy Subjects: A Double-Blind Cross-over Trial. *J. Nutr. Sci.* **2018**, *7*, e33. [CrossRef]

38. Santos-Hernández, M.; Miralles, B.; Amigo, L.; Recio, I. Intestinal Signaling of Proteins and Digestion-Derived Products Relevant to Satiety. *J. Agric. Food Chem.* **2018**, *66*, 10123–10131. [CrossRef]

39. Choi, S.; Lee, M.; Shiu, A.L.; Yo, S.J.; Hallden, G.; Aponte, G.W. GPR93 Activation by Protein Hydrolysate Induces CCK Transcription and Secretion in STC-1 Cells. *Am. J. Physiol. Gastrointest. Liver Physiol.* **2007**, *292*, G1366–G1375. [CrossRef]

40. Nakajima, S.; Hira, T.; Hara, H. Calcium-Sensing Receptor Mediates Dietary Peptide-Induced CCK Secretion in Enteroendocrine STC-1 Cells. *Mol. Nutr. Food Res.* **2012**, *56*, 753–760. [CrossRef]

41. Darcel, N.P.; Liou, A.P.; Tome, D.; Raybould, H.E. Activation of Vagal Afferents in the Rat Duodenum by Protein Digests Requires PepT1. *J. Nutr.* **2005**, *135*, 1491–1495. [CrossRef]

42. Caron, J.; Cudennec, B.; Domenger, D.; Belguesmia, Y.; Flahaut, C.; Kouach, M.; Lesage, J.; Goossens, J.-F.; Dhulster, P.; Ravallec, R. Simulated GI Digestion of Dietary Protein: Release of New Bioactive Peptides Involved in Gut Hormone Secretion. *Food Res. Int.* **2016**, *89*, 382–390. [CrossRef]

43. De Graaf, C.; Blom, W.A.; Smeets, P.A.; Stafleu, A.; Hendriks, H.F. Biomarkers of Satiation and Satiety. *Am. J. Clin. Nutr.* **2004**, *79*, 946–961. [CrossRef]

44. Domenger, D.; Caron, J.; Belguesmia, Y.; Lesage, J.; Dhulster, P.; Ravallec, R.; Cudennec, B. Bioactivities of Hemorphins Released from Bovine Haemoglobin Gastrointestinal Digestion: Dual Effects on Intestinal Hormones and DPP-IV Regulations. *J. Funct. Foods* **2017**, *36*, 9–17. [CrossRef]

45. Nishi, T.; Hara, H.; Asano, K.; Tomita, F. The Soybean β-Conglycinin β 51-63 Fragment Suppresses Appetite by Stimulating Cholecystokinin Release in Rats. *J. Nutr.* **2003**, *133*, 2537–2542. [CrossRef] [PubMed]

46. Osborne, S.; Chen, W.; Addepalli, R.; Colgrave, M.; Singh, T.; Tran, C.; Day, L. In Vitro Transport and Satiety of a Beta-Lactoglobulin Dipeptide and Beta-Casomorphin-7 and Its Metabolites. *Food Funct.* **2014**, *5*, 2706–2718. [CrossRef] [PubMed]

47. Tulipano, G.; Faggi, L.; Cacciamali, A.; Caroli, A.M. Whey Protein-Derived Peptide Sensing by Enteroendocrine Cells Compared with Osteoblast-like Cells: Role of Peptide Length and Peptide Composition, Focussing on Products of β-Lactoglobulin Hydrolysis. *Int. Dairy J.* **2017**, *72*, 55–62. [CrossRef]

48. Santos-Hernández, M.; Amigo, L.; Recio, I. Induction of CCK and GLP-1 Release in Enteroendocrine Cells by Egg White Peptides Generated during Gastrointestinal Digestion. *Food Chem.* **2020**, *329*, 127188. [CrossRef] [PubMed]

49. Wang, Y.; Chandra, R.; Samsa, L.A.; Gooch, B.; Fee, B.E.; Cook, J.M.; Vigna, S.R.; Grant, A.O.; Liddle, R.A. Amino Acids Stimulate Cholecystokinin Release through the Ca^{2+}-Sensing Receptor. *Am. J. Physiol. Gastrointest. Liver Physiol.* **2011**, *300*, G528–G537. [CrossRef] [PubMed]

50. Liou, A.P.; Sei, Y.; Zhao, X.; Feng, J.; Lu, X.; Thomas, C.; Pechhold, S.; Raybould, H.E.; Wank, S.A. The Extracellular Calcium-Sensing Receptor Is Required for Cholecystokinin Secretion in Response to l-Phenylalanine in Acutely Isolated Intestinal I Cells. *Am. J. Physiol. Gastrointest. Liver Physiol.* **2011**, *300*, G538–G546. [CrossRef]

51. Stagsted, J.; Zhou, J.; Jessen, R.; Torngaard Hansen, E. Dietary Peptides. U.S. Patent 16/062,429, 27 December 2018.

52. Le Nevé, B.; Daniel, H. Selected Tetrapeptides Lead to a GLP-1 Release from the Human Enteroendocrine Cell Line NCI-H716. *Regul. Pept.* **2011**, *167*, 14–20. [CrossRef]

53. Komatsu, Y.; Wada, Y.; Izumi, H.; Shimizu, T.; Takeda, Y.; Hira, T.; Hara, H. Casein Materials Show Different Digestion Patterns Using an in Vitro Gastrointestinal Model and Different Release of Glucagon-like Peptide-1 by Enteroendocrine GLUTag Cells. *Food Chem.* **2019**, *277*, 423–431. [CrossRef]

54. Diakogiannaki, E.; Pais, R.; Tolhurst, G.; Parker, H.E.; Horscroft, J.; Rauscher, B.; Zietek, T.; Daniel, H.; Gribble, F.M.; Reimann, F. Oligopeptides Stimulate Glucagon-like Peptide-1 Secretion in Mice through Proton-Coupled Uptake and the Calcium-Sensing Receptor. *Diabetologia* **2013**, *56*, 2688–2696. [CrossRef]

55. Liu, R.; Cheng, J.; Wu, H. Discovery of Food-Derived Dipeptidyl Peptidase IV Inhibitory Peptides: A Review. *Int. J. Mol. Sci.* **2019**, *20*, 463. [CrossRef] [PubMed]

56. Domenger, D.; Cudennec, B.; Kouach, M.; Touche, V.; Landry, C.; Lesage, J.; Gosselet, F.; Lestavel, S.; Goossens, J.-F.; Dhulster, P.; et al. Food-Derived Hemorphins Cross Intestinal and Blood–Brain Barriers In Vitro. *Front. Endocrinol.* **2018**, *9*. [CrossRef] [PubMed]

57. Lacroix, I.M.E.; Chen, X.-M.; Kitts, D.D.; Li-Chan, E.C.Y. Investigation into the Bioavailability of Milk Protein-Derived Peptides with Dipeptidyl-Peptidase IV Inhibitory Activity Using Caco-2 Cell Monolayers. *Food Funct.* **2017**, *8*, 701–709. [CrossRef] [PubMed]

58. Lundquist, P.; Artursson, P. Oral Absorption of Peptides and Nanoparticles across the Human Intestine: Opportunities, Limitations and Studies in Human Tissues. *Adv. Drug Deliv. Rev.* **2016**, *106*, 256–276. [CrossRef]

59. Shimizu, M. Food-Derived Peptides and Intestinal Functions. *Biofactors* **2004**, *21*, 43–47. [CrossRef]

60. Shimizu, M. Interaction between Food Substances and the Intestinal Epithelium. *Biosci. Biotechnol. Biochem.* **2010**, *74*, 232–241. [CrossRef]

61. De Cicco, M.; Mamone, G.; Di Stasio, L.; Ferranti, P.; Addeo, F.; Picariello, G. Hidden "Digestome": Current Analytical Approaches Provide Incomplete Peptide Inventories of Food Digests. *J. Agric. Food Chem.* **2019**, *67*, 7775–7782. [CrossRef]

 *molecules*

Article

Hydration and Barrier Potential of Cosmetic Matrices with Bee Products

Jana Pavlačková [1], Pavlína Egner [1], Roman Slavík [2], Pavel Mokrejš [3,*] and Robert Gál [4]

[1] Department of Lipids, Detergents and Cosmetics Technology, Faculty of Technology, Tomas Bata University in Zlín, Vavrečkova 275, 76001 Zlín, Czech Republic; pavlackova@utb.cz (J.P.); egner@utb.cz (P.E.)
[2] Prodejní místa, Alveare, Ltd., Štěpnická 1137, 68606 Uherské Hradiště, Czech Republic; slavik@post.cz
[3] Department of Polymer Engineering, Faculty of Technology, Tomas Bata University in Zlín, Vavrečkova 275, 76001 Zlín, Czech Republic
[4] Department of Food Technology, Faculty of Technology, Tomas Bata University in Zlín, Vavrečkova 275, 76001 Zlín, Czech Republic; gal@utb.cz
* Correspondence: mokrejs@utb.cz; Tel.: +420-576-031-230

Academic Editors: María Dolores Torres and Elena Falqué López
Received: 20 April 2020; Accepted: 26 May 2020; Published: 28 May 2020

Abstract: Honey, honey extracts, and bee products belong to traditionally used bioactive molecules in many areas. The aim of the study was primarily to evaluate the effect of cosmetic matrices containing honey and bee products on the skin. The study is complemented by a questionnaire survey on the knowledge and awareness of the effects and potential uses of bee products. The effect of bee molecules at various concentrations was observed by applying 12 formulations to the skin of the volar side of the forearm by non-invasive bioengineering methods on a set of 24 volunteers for 48 h. Very good moisturizing properties have been found in matrices with the glycerin extract of honey. Matrices containing forest honey had better moisturizing effects than those containing flower honey. Barrier properties were enhanced by gradual absorption, especially in formulations with both glycerin and aqueous honey extract. The observed organoleptic properties of the matrices assessed by sensory analysis through 12 evaluators did not show statistically significant differences except for color and spreadability. There are differences in the ability to hydrate the skin, reduce the loss of epidermal water, and affect the pH of the skin surface, including the organoleptic properties between honey and bee product matrices according to their type and concentration.

Keywords: bee products; bioactive molecules; cosmetics; emulsion; functional matrices; honey; hydration; organoleptic properties; transepidermal water loss

1. Introduction

Bees are important to humans not only by pollinating many species of plants, thereby helping to multiply them, but also by creating unique products containing a range of bioactive molecules that have been used for centuries in human nutrition, folk medicine, pharmacy, and cosmetics. Bee products may be of both vegetal and animal origin. The products of vegetal origin include those molecules that bees collect in the wild and bring to the hive, where they process them for their needs. These include propolis—the compound from the flower and leaf buds of alder, birch and other trees, pollen—the floral pollen of flowering plants and honey—the nectar of flowering plants, or honeydew produced by the *Homoptera insecta*. Bee products of animal origin include substances that the bee itself secrete in its own body. It is bee venom—secretion of the venom gland, royal jelly —secretion of the pharyngeal and mandible glands of worker bees, and beeswax—the secretion of wax-forming glands of worker bees [1,2]. The composition, color, aroma, and biological activity of bee products depend on the location, time, and source of the plant from which they are obtained.

Among the raw materials for cosmetic matrices, honey is the most important. "Honey" or "Mel" is included in the International Nomenclature of Cosmetic Ingredients (INCI) as an moisturizing/humectant/emollient product [3]. Honey is useful in products of skin care, and its regular application contributes to the skin juvenility and the reduction of wrinkle formation [4]. Honey is used in variable proportions according to the type of cosmetic. Generally, a lower concentration (0.5–5%) is used for foams, creams and emulsions, while a higher concentration (10–15%) is used for anhydrous ointments. Honey is most commonly used in the range of 1% to 10% in products such as lip ointments, cleansing milk, hydrating creams or gels, after-sun products, tonic lotions, shampoos, and conditioners. Higher concentrations (up to 70%) can be used for combinations of honey with oils, gelling agents, and emulsifiers or in face masks [5,6]. Honey is also used as an alternative to traditional emulsifiers in body lotions for bathing and shampooing, where they make up 50% to 50% surfactants [7].

Beeswax is used to modify the texture of cosmetic products: 1–3% for creams and ointments, balms, and lotions, 6–12% for mascara, and 6–20 % for eye shadow. It is included in deodorants (up to 35%), depilatory preparations (up to 50%), hair cosmetics (1–10%), lipsticks (10–15%), and other products. Its presence also improves the stability of the formulations. Another bee product used primarily for its antimicrobial effects is pollen, which is added to dry shampoos, creams, and tonics [8]. For its anti-aging effect, a royal jelly is widely used. Royal jelly extract increases the natural moisturizing factor. The addition of 0.05% to 1.0% stimulates and nourishes the epidermis; it is used e.g., in face lotions, body milk, hair cosmetics, and soaps [9]. Propolis has regenerative, antioxidant, anti-inflammatory, antiseptic, antifungal, bacteriostatic, astringent, antispasmodic, anesthetic, anticancer and photoprotective effects [10,11]. It is added in a concentration of 1% to 2% to aftershave, after-bath and oral care products, shampoos, deodorants, soaps, and creams [12].

The development of bee products for dermal applications may take different directions in the future. Burlando and Cornara [7] see one way in ethnopharmaceuticals surveys focused on significant biological properties in the extraordinary variety of mono- and polyfloral honeys. Another possibility is to carry out chemical and biological research focused on the chemical composition of honey and its pharmacological efficacy, thus opening the way to new medical procedures supporting human health [13].

The use of honey, honey extracts, and bee products as bioactive molecules in many cosmetic products (especially body and hair cosmetics) has been known for some time. A variety of publications describe the diverse effect of these formulations; however, none of them quantitates the moisturizing and barrier effect containing various bee products. The aim of the present paper is to assess (1) the moisturizing and barrier properties of emulsion matrices with the addition of honey and bee products on the skin; (2) perform sensory evaluation of the emulsion matrices with the addition of honey and bee products; and ascertain (3) by means of a questionnaire survey the effects of cosmetic matrices with the addition of honey and bee products, the reasons and frequency of use of honey cosmetics. The specific hypotheses tested are as follows. (1) Higher skin hydration may be expected for emulsion matrices with a higher content of honey or bee products. (2) Emulsion matrices with a higher honey and wax content may be expected to have a difference in their spreadability. (3) Honey and/or bee products are expected to be popular with female respondents.

2. Results and Discussion

2.1. Biophysical Characteristics

The measurement of hydration, barrier, and pH effects on the skin due to the effect of prepared cosmetic matrices were preceded by a 0.5% sodium lauryl sulfate (SLS) skin pretreatment, the so-called washing test, which simulated the use of cosmetics during personal hygiene, such as showers or washes, and another purpose was to eliminate differences in skin properties at the site of intended application. The skin pretreatment is presented in the Figures 1–3, characterizing the measured parameters by a continuous horizontal line corresponding to the average value of the monitored parameter measured before the emulsion application. This method of pretreating the skin with SLS solution in various concentrations is referred to in a number of works devoted to testing various preparations [14–16].

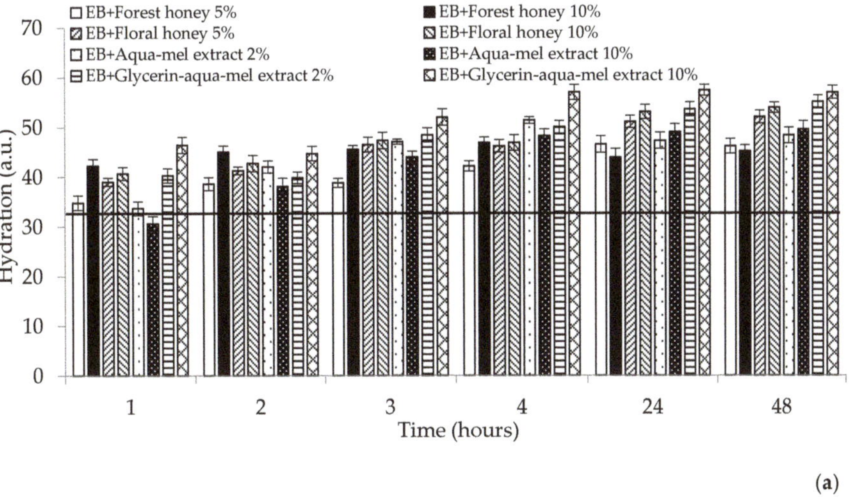

(a)

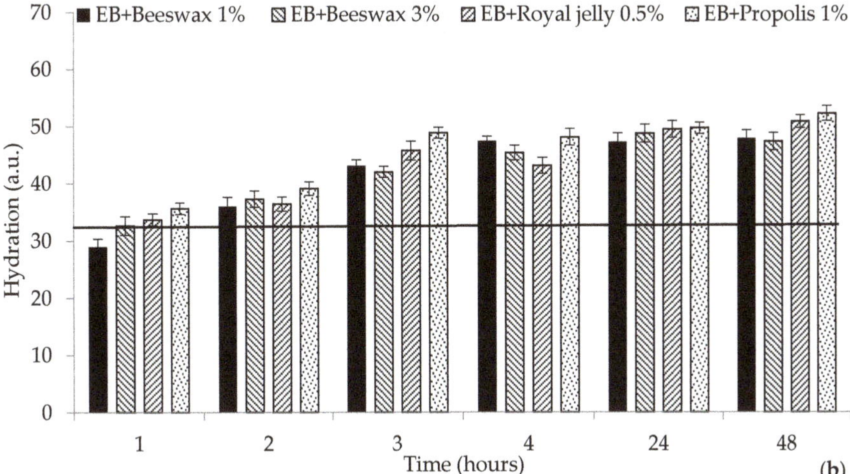

(b)

Figure 1. Hydration effect of tested cosmetic matrices after sodium lauryl sulfate (SLS) pretreatment (continuous horizontal line) over the studied period: (**a**) honey and honey extracts (**b**), other bee products. EB: emulsion base.

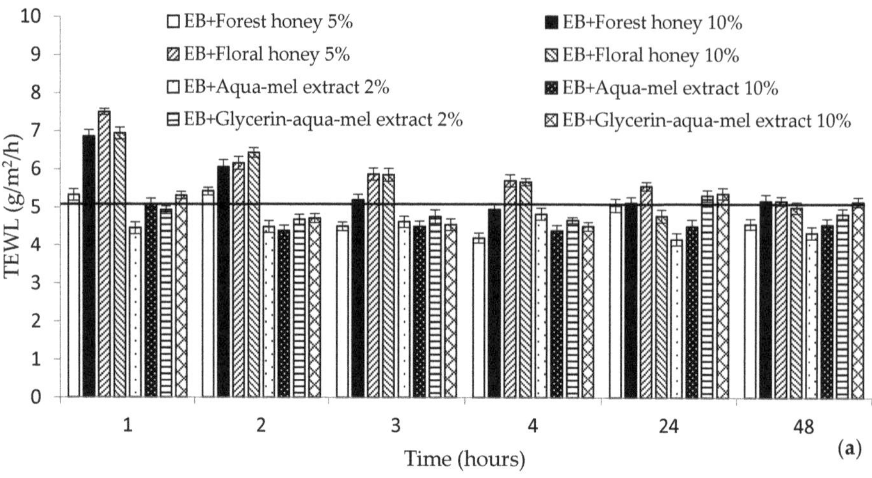

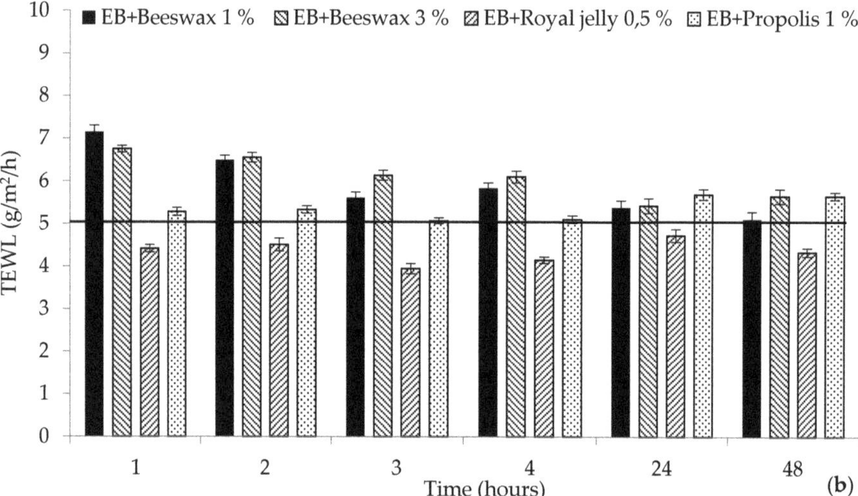

Figure 2. Transepidermal water loss (TEWL) after SLS pretreatment (continuous horizontal line) and application of the tested cosmetic matrices over the studied period: (**a**) honey and honey extracts (**b**), other bee products. EB: emulsion base.

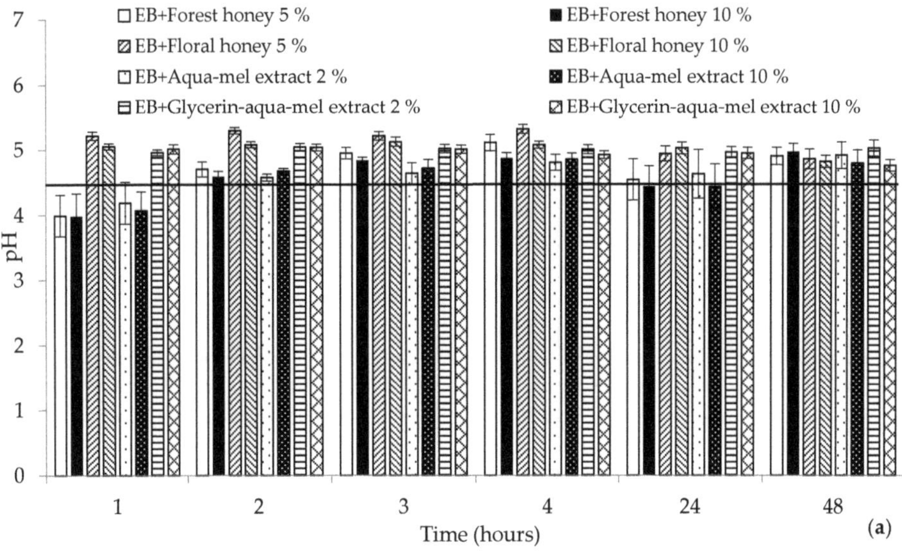

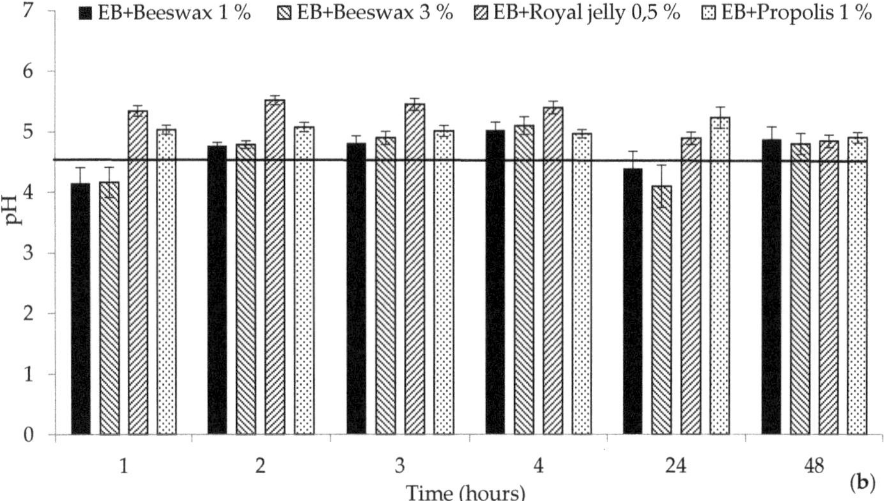

Figure 3. Values for pH after SLS pretreatment (continuous horizontal line) and application of the tested cosmetic matrices over the studied period: (**a**) honey and honey extracts (**b**), other bee products. EB: emulsion base.

The hydrating ability of emulsion matrices with honey, honey extracts, and other bee products is shown in Figure 1. Emulsion matrices with glycerin extract of honey have probably shown a synergistic hydrating effect of the ingredients contained in the extract; the final increase in hydration after degreasing the skin treated with both formulations was about 60%. Emulsions containing higher concentrations of bee products hydrated the skin more efficiently except for the aqueous extract formulation, where, on the contrary, the lower concentrations proved to be preferable after four hours of exposure. An interesting difference was observed during the treatment of forest and flower honey formulations, with flower honey showing a higher hydrating activity of approximately 15% after 24 h, which can be attributed to a slightly higher fructose and glucose content in flower honey (see Table 1). This is consistent with the results of Jiménez et al., which state that active hydration is influenced by the content of sugars, mainly fructose and glucose [17]. These sugars form hydrogen bridges with water and maintain the moisture of the skin horny layer. This creates a protective non-greasy film on the skin to help maintain water in the skin [18]. Burlando and Cornara [7] in their review extend this knowledge to the influence of other substances present such as amino acids and organic acids, which can supplement the natural moisturizing factors of the horny layer. It is known that the biological properties of a certain type of honey are determined by the nectar-producing plants; therefore, botanical resources are of great importance in cosmetics [19]. Honey with a higher content of better soluble fructose with respect to glucose is recommended as being more suitable for cosmetic products because of the lower risk of their crystalline form. There is a study [20] in which volunteers observe the hydrating effect of honey-containing emulsions from different bee species, where the *Apis mellifera* honey formulation proved to be the most hydrating active after two hours of application, which was followed by preparations containing the honey of *Melipona fasciculata* and the honey of *Tetragonula carbonaria* with the least hydrating effect. The skin treated with other bee product formulations also achieved a higher proportion of water in its corner layer. Hydrating effects of bee products are mentioned in publication [21]; the samples containing a higher amount of beeswax were absorbed into the skin more slowly than other tested formulations. The hydration of stratum corneum is crucial for the integrity and regulating the barrier properties.

Table 1. Composition of forest and flower honey.

Component	Forest Honey (% ± SD)	Flower Honey (% ± SD)
Fructose	31.30 ± 1.95	38.40 ± 2.01
Glucose	36.80 ± 4.60	33.20 ± 4.08
Sucrose	2.30 ± 0.59	2.90 ± 0.55
Maltose	5.50 ± 0.48	3.10 ± 0.34
Oligosaccharides	3.70 ± 0.61	2.20 ± 0.41
Moisture	18.60 ± 2.17	18.20 ± 2.04
Enzymes	0.16 ± 0.02	0.56 ± 0.03
Vitamins	0.11 ± 0.02	0.21 ± 0.02
Ash	1.04 ± 0.04	0.45 ± 0.03
Free acids	0.15 ± 0.02	0.35 ± 0.03
Amino acids	0.34 ± 0.02	0.43 ± 0.02

The effect of emulsion matrices with honey, honey extracts, and other bee products on the skin barrier function is shown in Figure 2. The highest loss of epidermal water (TEWL) was monitored after treatment of the skin with an SLS solution that disrupted the skin barrier. This mechanism of increasing TEWL after treatment with diversely concentrated SLS solution is described by Tupker et al. [22]. Some authors [23] recommend replacing the SLS treatment with anionic phosphorus derivatives of alkyl polyglucsides that are more gentle to the skin barrier. Park and Eun [24] studied the concentration series of SLS solution, and the increase in TEWL was statistically significant relative to the higher concentration of the solution. The reduction of water loss from the epidermis proceeded gradually over the observed time intervals with the gradual absorption of emulsions. The higher increase in TEWL does not always correspond to higher concentrations of the studied active substances in cosmetic matrices. The reduction of dermal water loss was proved after the treatment of the skin with both glycerin and aqua mel extract during the 4 h of the experiment. In the study [25], the effects of aqua solution of glycerin in the concentration range 1–10% on the skin of women volunteers pretreated with 10% SLS solution under occlusion for 3 h was examined. Even at 2% concentration of glycerin, the water-holding capacity was enhanced. When evaluating matrices containing flower and forest honey, less water loss was observed after the treatment of the skin with a formulation with the addition of forest honey. In a randomized controlled trial by Zahmatkesh et al. [26], a mixture of olive oil, sesame oil, and honey was demonstrated to be a useful treatment for burns, by preventing infections, accelerating tissue repair, and facilitating debridement. Among other bee products proven to be suitable were emulsion formulations with wax and royal jelly, which have been shown to prevent the dehydration of the skin.

The last observed parameter affected by the matrices prepared was skin pH. Figure 3 shows a pH shift to the neutral area after application of the tested matrices, most notably for honey and beeswax formulations. Tested cosmetic matrices do not disturb the natural pH of the skin. The skin surface is naturally slightly acidic; the pH ranges from 4.0 to 5.5 depending on the location [27]. The skin pretreatment by degreasing represented a potential risk of removing dermal lipids and bacterial flora, to which Seweryn el al. [28] draw attention. From the pH values after the topical application of matrices containing honey and bee products, it can be concluded that no disruption of natural acid skin mantle was detected, which enables an indirect prediction of the influence of studied matrices on the skin. There was no irritating reaction (assessed visually) on the skin treated with the tested matrices, although honey-based cosmetics and cosmetics made of other bee products can function as sensitizers, as described in some publications [29,30]. To conclude, from a physiobiological approach, a casual dry skin may be treated with honey and other bee products matrices giving good efficacy.

2.2. Sensory Analysis

There were no statistically significant differences between the cosmetic matrices of the honey and bee products formulations assessed by the ranking test evaluating overall sample preference. Statistically significant differences at the 95% level of significance were found in the ranking test assessing the color of the matrices between samples AG, CF, CG, CE, CH, BG, BE, BH, DA, DF, DG, DE, and DH; designation of the samples is described in Chapter 3.2. No statistically significant differences ($p < 0.05$) were found in the paired comparative test observing the pleasantness of emulsion matrices with both 5% and 10% flower and forest honey content. Furthermore, the results of a paired test comparing the spreadability of the formulations with 1% and 3% beeswax content were also statistically significant, in favor of a sample with a lower beeswax content. Additions of beeswax can modify the texture and viscosity of the matrices, but also act as a smoothing and opacifying agent [19].

2.3. Questionnairre Survey

The questionnaire survey revealed that honey was used by 92% of the respondents, propolis was used by 40% of the respondents, royal jelly was used by 14% of the respondents, wax was used by 35% of the respondents, and bee venom was used by 3% of the respondents. As a reason for using honey cosmetics, 15% interviewed accepted the advice of a friend, 5% decided on the basis of information from the media, and 3% of them stated physician recommendations. As a result of skin problems, 5% searched for bee products, and 61% of those interviewed stated other reasons. Most (88%) women were aware of the healing effects of honey and bee products; meanwhile, 9% of the respondents suffered from allergy to bee products, of which 5% were to bee venom, 3% were to pollen, and 1% were to honey. The processing of honey and bee products was carried out by 4% of respondents. More than half of the respondents had experience with cosmetics containing honey and bee products. Cosmetics with honey and bee products were used by 55% of women. The reason was that these cosmetic products are used to improve psoriasis and dry skin condition (45%), improve the wound state (25%), strengthen immunity (12%), reduce acne (6%), reduce inflammation (6%), reduce eczema (3%), and relieve pain (3%). The cosmetic products were most often applied to the body, face, hands, eye area, hair, and oral cavity. The most common forms of cosmetic products were toothpastes, creams, ointments, balms, tinctures, lotions, masks, mouthwashes, gels, and shampoos.

3. Materials and Methods

3.1. Analysis of Forest and Flower Honey

The analysis of forest and flower honey was carried out at the Institute of Environment of the Faculty of Technology (Zlín, Czech Republic) in cooperation with the Slovak Academy of Sciences (Bratislava, Slovak Republic) see Table 1. Moisture was determined by refractometric method according to DIN 10752 [31]. The principle is that the water content is determined from the refractive index of honey (which increases with the dry matter content). The acidity was determined by titrating a sample of honey dissolved in 0.1 mol/L NaOH according to Lord et al.; acidity expresses the amount of all free acids in honey [32]. The mineral content was determined gravimetrically after burning and annealing the sample at 600 °C [33]. Amino acids were determined by gas chromatography with flame ionization and mass spectrometric detection (Shimadzu, Tokyo, Japan) [34,35]. The content of reducing saccharides (glucose, fructose, maltose), non-reducing disaccharide (sucrose) and oligosaccharides was determined by high-performance liquid chromatography with a refractive index detector (Shimadzu, Tokyo, Japan) [36–38]. Water-soluble vitamins (B2, B3, B5, B9, and C) were determined by reversed phase high-performance liquid chromatography [39]. Enzyme (diastase) was determined according to methods described by Edwards et al. [40] and Hadorn [41].

3.2. Procedure of Preparation of Cosmetics Matrices

A total of 12 emulsion matrices containing *Apis mellifera* European honey and bee products were prepared in various concentrations, selected according to available sources [7,12,42–44] and specifications from manufacturers [45,46]. The aim of the study was to test the effectiveness of recommended minimal and maximal additions of honey and bee products into an emulsion base (EB). To prepare the matrices, the EB (Fagron, Czech Republic) with following composition (according to International Nomenclature of Cosmetic Ingredients) was used: *Aqua, Paraffin, Paraffinum Liquidum, Cetearyl Alcohol, Laureth 4, Sodium Hydroxide, Carbomer, Methylparaben, Propylpareben*. Part of the honey whose skin effect was studied came from colonies located on the roof of the Faculty of Technology. The list of formulations is as follows. EB with the addition of flower honey (Tomas Bata University in Zlín, Czech Republic) in concentrations of 5% and 10% (samples A, B) and forest honey (Hostyn-Vsetin Highlands, 25 km far from Tomas Bata University in Zlín, Czech Republic) in concentrations of 5% and 10% (samples C, D). Other products added to the basic formulation were glycerin–aqua–mel extract PHYTAMI® HONEY-F06 (Alban Mueller International, France) at concentrations of 2% and 10% (samples E, F), aqua–mel extract CRODAROM HONEY (Crodarom, France) at concentrations of 2% and 10% (samples G, H), beeswax (Včelpo, Czech Republic) at concentrations of 1% and 3% (samples I, J), royal jelly (Včelpo, Czech Republic) at a concentration of 0.5% (sample K), and propolis (Včelpo, Czech Republic) at a concentration of 1% (sample L).

The procedure for preparing the matrices was chosen with respect to the particular honey or bee products. Flower honey, forest honey, and royal jelly were homogenized into EB on a Heidolph stirrer at 2000 rpm for 10 min at room temperature. Propolis and beeswax had to be treated before mixing into EB. An ethanolic tincture was formed from the crude propolis by mixing 1 part propolis (100 mL) and 2 parts of 96% ethanol (200 mL). This mixture was left to infuse with occasional shaking for 7 days and then filtered and added to EB. Beeswax creams were prepared by weighing the required amount of wax and EB separately. The wax was heated in a water bath at 70 °C for 10 min, and the heated wax was added to EB at 60 °C. This was followed by homogenization with a Heidolph stirrer at 2000 rpm at laboratory temperature.

3.3. Instrumental Techniques and Study Design

In the field of experimental dermatology and cosmetology, non-invasive methods are widely used, which allow quantitative evaluation of parameters describing the barrier function of the skin. These methods were also applied in this study. The CORNEOMETR® CM 825 corneometric probe

(Courage & Kazaka Electronic, Cologne, Germany) was used to measure the water content of the *stratum corneum*, based on the evaluation of changes in electrical capacity on the skin surface, using relatively high dielectric water constants. The results are displayed in arbitrary units (a.u.). Another parameter measured was transepidermal water loss (TEWL), which was monitored by a TEWAMETER® TM 300 probe (Courage & Kazaka Electronic, Cologne, Germany). In principle, the flow of water vapor above the *stratum corneum* into the open chamber of a cylindrical shape with two pairs of sensors for temperature and relative humidity is determined. TEWL is calculated from the difference between the two measurement points using Fick's law of diffusion and displayed in grams per square meter per hour ($g/m^2/h$). Skin-pH-meter® PH 905 (Courage & Kazaka Electronic, Cologne, Germany) was used to determine skin acidity. The specially designed probe consists of a flat-topped glass electrode for full skin contact, which was connected to a voltmeter. The system measures potential changes due to the activity of hydrogen cations surrounding the very thin layer of semisolid forms at the top of the probe. The changes in voltage are displayed as pH.

Measurements were performed on 24 healthy women (aged 23 to 49 years, mean age 36 years) with no history of atopic eczema or other skin diseases. The volunteers were divided into two groups, each testing six formulations (see the study design in Table 2). Volunteers were acquainted in advance with the purpose and course of the measurement. Informed consent was obtained from all of them, and said study was approved by the International Ethical Guidelines for Health-Related Research Involving Humans [47]. For 12 h prior to and during the study, the volunteers were not allowed to apply any topical cosmetic products; only an evening shower water was permitted. Measurements were carried out in an air-conditioned room (temperature 22.0–24.0 °C, relative humidity 45.0–50.0%). All measurements were performed after a rest of 20 min for equilibration. The volar side of the forearm of the right and left hand was divided into test sites with an area of 8 cm^2 (see design in Table 2), which were pretreated for 4 h with 0.5% SLS solution (Sigma-Aldrich, Czech Republic) prepared in saline. After this pretreatment, the following indicators were measured on each test site: hydration with a corneometric probe, TEWL using a tewameter, and acidity of the skin by a pH probe. The untreated spot, designated as a control, was used to compare any irritant reactions to the skin. Then, 0.5% SLS solution and honey and bee products matrices were applied to each site. The effect of the applied samples on the *stratum corneum* was monitored in all volunteers after 1, 2, 3, 4, 24, and 48 h in the same order as after SLS treatment. Hydration was measured five times at each test site. TEWL measurements were performed 15 times at each test site. Since this is dependent on skin temperature, ambient temperature, and the temperature of a probe itself, the first five values were eliminated.

Table 2. Design of the volar side of forearms of two groups of volunteers with tested formulations; concentrations of honey, honey extracts, and other bee products are in EB (*w/w*).

Volar Side of the Left Forearm		Volar Side of the Right Forearm	
1st group of volunteers	2nd group of volunteers	1st group of volunteers	2nd group of volunteers
Control		Aqua-mel extract 2%	Floral honey 10%
SLS		Aqua-mel extract 10%	Glycerin–aqua–mel extract 2%
Forest honey 5%	Royal jelly 0.5%	Beeswax 1%	Glycerin–aqua–mel extract 10%
Forest honey 10%	Floral honey 5%	Beeswax 3%	Propolis 1%

3.4. Sensory Analysis and Questionnairre Survey

The panel of assessors consisted of a total of 12 assessors at a trained assessor level. The assessors were acquainted with the objective of analysis and instructed on the way of evaluation of samples of individual products. The sensory evaluation and equipment of the sensory laboratory were in compliance with regulations defined by International Standards ISO 6658 [48] and ISO 8589 [49].

The temperature in the room was at 20.0–22.0 °C, relative humidity 45.0–50.0% under conditions of artificial light. The sensory analysis included ordinal tests focused on overall sample preference and sample color preference for 8 selected formulations: containing 5% and 10% flower honey (samples A and B), 5% and 10% forest honey (samples C and D), 2% and 10% glycerin–aqua–mel extract (samples E and F), 2% and 10% aqua–mel extract (samples G and H). Then, paired comparative tests for honey and beeswax formulations were included. The first paired test examined the comfort of the 5% flower and forest honey samples on the skin. Another paired test evaluated a pair of samples with 10% content of flower and forest honey in the same characteristics. The last paired test evaluated the spreadability of cream samples containing 1% and 3% beeswax.

A questionnaire survey was proposed to ascertain data for the use of honey and bee products in cosmetics and to map the existence of knowledge about the effects of these products. The questionnaire included socio-demographic questions determining age (15–20 years, 21–30 years, 31–40 years, 41–50 years, 51–60 years, 61–70 years), place of residence (village, town, city), education (elementary, apprenticeship, apprenticeship with school-leaving exam, secondary, higher professional and university education) and field of the employment of respondents: health service, education, food industry, services (cosmetic, wellness, hairdressing), nutrition consultancy, state administration, and others. This was followed by questions focused on honey and bee products, which are presented in Table 3. The survey was anonymous, the target group of respondents were women. In total, 120 questionnaires were distributed with a return of 83%. The proportion of respondents in individual age categories was even except for respondents aged 61 to 70, which was only 5%. Most respondents had secondary and university education (43% and 51%).

Table 3. Summary of questions and answers possibilities in the questionnaire.

No.	Question
1	Which products do you use? Answer possibility: honey, propolis, royal jelly, wax, bee venom, pollen
2	What reasons led you to use honey and bee products? Answer possibility: physician recommendation, friend's recommendation, advertising, skin problems, others
3	Do you think bee products has healing effects? Answer possibility: yes/no/don't know
4	Are you allergic to any of the listed products: honey, propolis, royal jelly, wax, bee venom, pollen? Answer possibility: yes/no
5	Do you process honey or bee products for the production of ointments, tinctures, emulsions, etc. for home use? Answer possibility: yes/no
6	Do you use honey cosmetics? Answer possibility:/no
(a)	What is the reason for using honey cosmetics? Answer possibility: pain relief, wound healing, acne, inflammation, eczema, psoriasis, dry skin, strengthening immunity
(b)	What types of honey cosmetics do you use? Answer possibility: healing cosmetics (drops/tincture/spray/gel/ointment/balm/cream), cleaning cosmetics (wipes/water/milk/peeling/mask/toothpaste/mouthwash), cosmetics for men (shampoo/after shave/men's cream), body cosmetics (milk/balm/bath foam/soap/shower shampoo), hair cosmetic (shampoo/hair lotion/balm)

3.5. Statistical Analysis

Analyses of forest and flower honey were performed in triplicate; the arithmetic mean and standard deviation values were calculated using Microsoft Office Excel 2013 (Microsoft, Santa Rosa, California, CA, USA). Hydration, TEWL, and pH values were recorded and processed via MPA 5 station operating software (Courage & Khazaka Electronic, Cologne, Germany). The results of biophysical characteristics reported as the mean values with standard deviations were carried out in Excell software (version 10, Microsoft, Santa Rosa, California, CA, USA). The results of sensory analysis—ordinal and pair tests—were evaluated by Friedman's test at 5% significance level. Frequency expression was used to evaluate the data obtained by the questionnaire survey.

4. Conclusions

Between selected honey and bee products, there are differences in the ability to hydrate the skin and improve its barrier properties, including adjusting the acidity of the skin surface. Their effectiveness is dependent on the type and concentration of the product incorporated into the cosmetic vehicle. Very good moisturizing properties have been found in emulsion matrices with a glycerin extract of honey, which is attributed to the synergistic effect of glycerin present, which is a traditional humectant very often used in cosmetic products. Cosmetic matricess containing higher concentrations of honey or bee products hydrated the skin more effectively except for the aqua–mel extract formulation where lower concentrations were found to be more favorable. A very interesting difference in the ability to hydrate was observed in forest and flower honey formulations, where forest honey had a higher hydrating activity. Even the skin treated with matrices containing other bee products had a higher proportion of water in its corner layer. Barrier properties were enhanced by gradual absorption, especially in samples with both glycerin and aqueous honey extract. A favorable finding from the measurement results was the shift of the skin pH to the neutral area. By sensory analysis, differences in color of emulsion matrices were evaluated from organoleptic properties. A paired test found a difference in the spreadability of formulations with different amounts of beeswax. The reasons and frequency of using of cosmetics containing honey and bee products vary. Summarizing the information obtained, honey and bee products as substances of natural origin are very popular and traditionally used primarily for their unique effects not only in cosmetics, but also in many other areas. Cosmetic matrices enriched with honey or bee products are suitable for the care of skin which is repeatedly exposed to surfactants contained in cosmetic personal care products and in variety of cleansing agents.

Author Contributions: Conceptualization, J.P.; methodology, J.P.; software, P.E.; validation, J.P.; formal analysis, R.S.; investigation, P.E.; resources, R.S.; data curation, J.P. and P.E.; writing—original draft preparation, J.P. and P.M.; writing—review and editing, P.M.; visualization, R.S.; supervision, P.M.; project administration, R.G. All authors have read and agreed to the published version of the manuscript.

Funding: This research was funded by the Internal strategic project of Tomas Bata University in Zlín focused on recycling technologies for natural and synthetic polymers and utilization of products obtained from them.

Acknowledgments: The authors thank to Marie Dvořáková (Zlín, Czech Republic) for assistance during laboratory experiments and to David Dohnal (Olomouc, Czech Republic) for editing of the manuscript.

Conflicts of Interest: The authors declare no conflict of interest.

References

1. Hajdušová, J. *Bee Products from the Physician Sight [in Czech]*, 1st ed.; Český svaz včelařů: Praha, Czech Republic, 2006; pp. 10–45.
2. Přidal, A. *Bee Products [in Czech]*, 1st ed.; MZLU: Brno, Czech Republic, 2005; pp. 48–80.
3. Gottschlack, T.E.; McEwen, G.N. *International Cosmetic Ingredient Dictionary and Handbook*, 13th ed.; The Personal Care Products Council: Washington, DC, USA, 2010.
4. Crane, E. *A Book of Honey*; Oxford University Press: Oxford, UK, 1980; pp. 39–63.
5. Nilforoushzadech, M.A.; Amirkhani, M.A.; Zarrintaj, P.; Salehi Moghaddam, A.; Mehrabi, T.; Alavi, S.; Mollapour, S.M. Skin care and rejuvenation by cosmeceutical facial mask. *J. Cosmet. Dermatol.* **2018**, *17*, 693–702. [CrossRef] [PubMed]
6. McLoone, P.; Warnock, M.; Fyfe, L. Honey: A realistic antimicrobial for disorders of the skin. *J. Microbial. Immunol. Infect.* **2016**, *49*, 161–167. [CrossRef] [PubMed]
7. Burlando, B.; Cornara, L. Honey in dermatology and skin care: A review. *J. Cosmet. Dermatol.* **2013**, *12*, 306–313. [CrossRef] [PubMed]
8. Gordon, L.A. Compositae dermatitis. *Australas J. Derm.* **1999**, *40*, 123–130. [CrossRef] [PubMed]

9. Bogdanov, S. Royal Jelly, Bee Brood: Composition, Health, Medicine: A Review. Available online: https://www.bee-hexagon.net/files/file/fileE/Health/RJBookReview.pdf (accessed on 15 September 2019).

10. Saewan, N.; Jimtaisong, A. Natural products as photoprotection. *J. Cosmet. Dermatol.* **2015**, *14*, 47–63. [CrossRef]

11. Banskota, A.H.; Tezuka, Y.; Kadota, S. Recent progress in pharmacological research of propolis. *Phytother. Res.* **2001**, *15*, 561–571. [CrossRef]

12. Krell, L. Value-Added Products from Beekeeping. Available online: http://www.fao.org/3/w0076e/w0076e00.htm (accessed on 16 September 2019).

13. Meo, S.A.; Al-Asiri, S.A.; Mahesar, A.L.; Ansari, M.J. Role of honey in modern medicine. *Saudi J. Biol. Sci.* **2017**, *24*, 975–978. [CrossRef]

14. De Paepe, K.; Derde, M.P.; Roseeuw, D.; Rogiers, V. Claim substantion and efficiency of hydrating body lotions and protective creams. *Contact Dermat.* **2000**, *42*, 227–234. [CrossRef]

15. Lodén, M.; Wessman, W. The influence of a cream containing 20% glycerin and its vehicle on skin barrier properties. *Int. J. Cosmet. Sci.* **2001**, *23*, 115–119. [CrossRef]

16. Polaskova, J.; Pavlackova, J.; Egner, P. Effect of vehicle on the performance of active moisturizing substances. *Skin Res. Technol.* **2015**, *21*, 403–412. [CrossRef]

17. Jiménez, M.M.; Fresno, M.J.; Sellés, E. The galenic behaviour of a dermopharmaceutical excipient containing honey. *Int. J. Cosmet. Sci.* **1994**, *16*, 211–226. [CrossRef] [PubMed]

18. Ahshawat, M.S.; Saraf, S.; Saraf, S. Preparation and characterization of herbal creams for improvement of skin viscoelastic properties. *Int. J. Cosmet. Sci.* **2008**, *30*, 183–193. [CrossRef] [PubMed]

19. Ediriweera, E.R.H.S.S.; Premarathna, N.Y.S. Medicinal and cosmetic uses of bee's honey–A review. *Ayu* **2012**, *33*, 178–182. [CrossRef]

20. Isla, M.; Cordero, A.; Díaz, L.; Pérez-Pérez, E.M.; Vit, P. Cosmetic properties of honey. 1. Antioxidant activity. In *Stingless Bees Process Honey and Pollen in Cerumen Pots*; Vit, P., Roubik, D.W., Eds.; Universidad de los Andes: Mérida, Venezuela, 2013; pp. 1–8.

21. Márquez, R.; Bálsamo, S.; Morales, F.; Ruiz, N.; García, A.; León, R.; Montes, A.; Nava, N.; Noguera, Y.; Quintero, A.; et al. Technological use of beeswax for obtaining organic products, non-toxic for the human being. *Revista Ciencia e Ingeniería* **2019**, *40*, 17–26.

22. Tupker, R.A.; Willis, C.; Berardesca, E.; Lee, C.H.; Fartasch, M.; Agner, T.; Serup, J. Guidelines on sodium lauryl sulfate (SLS) exposure tests. A report from the Standardization Group of the European Society of Contact Dermatitis. *Contact Dermat.* **1997**, *37*, 53–69. [CrossRef] [PubMed]

23. Seweryn, A.; Bujak, T. Application of Anionic Phosphorus Derivatives of Alkyl Polyglucosides for the Production of Sustainable and Mild Body Wash Cosmetics. *ACS Sustain. Chem. Eng.* **2018**, *6*, 17294–17301. [CrossRef]

24. Park, K.B.; Eun, H.C. A study of skin responses to follow-up, rechallenge and combined effects of irritants using non-invasive measurements. *J. Derm. Sci.* **1995**, *10*, 159–165. [CrossRef]

25. Atrux-Tallau, N.; Romagny, C.; Padois, K.; Denis, A.; Haftek, M.; Falson, F.; Pirot, F.; Maibach, H.I. Effects of glycerol on human skin damaged by acute sodium lauryl sulphate treatment. *Arch. Derm. Res.* **2010**, *302*, 435–441. [CrossRef]

26. Zahmatkesh, M.; Manesh, M.J.; Babashahabi, R. Effect of olea ointment and acetate mafenide on burn wounds–a randomized clinical trial. *Iran. J. Nurs. Midwifery Res.* **2015**, *20*, 599–603. [PubMed]

27. Polášková, S. Year-round care of a child's skin [in Czech]. *Pediatr. Praxi* **2012**, *13*, 96–100.

28. Seweryn, A.; Klimaszewska, E.; Ogorzałek, M. Improvement in the Safety of Use of Hand Dishwashing Liquids through the Addition of Sulfonic Derivatives of Alkyl Polyglucosides. *J. Surfact. Deterg.* **2019**, *22*, 743–750. [CrossRef]

29. Kiistala, R.; Hannuksela, M.; Mäkinen-Kiljunen, S.; Niinimäki, A.; Haahtela, T. Honey allergy is rare in patients sensitive to pollens. *Allergy* **1995**, *50*, 844–847. [CrossRef] [PubMed]

30. Rosmilah, M.; Shahnaz, M.; Patel, G.; Lock, J.; Rahman, D.; Masita, A.; Noormalin, A. Characterization of major allergens of royal jelly Apis mellifera. *Trop Biomed.* **2008**, *25*, 243–251. [PubMed]

31. Deutsche Industrie Norm. Bestimmung des Wassergehalts in Honig. DIN 10752. 1992.

32. Lord, D.W.; Scotter, M.J.; Whittaker, A.D.; Wood, R. The determination of acidity, apparent reducing sugar and sucrose, hydroxymethylfurfural, mineral, moisture, water-insoluble solids contents in honey; collaborative study. *J. Assoc. Publ. Anal.* **1988**, *26*, 51–76.

33. Akharaiyi, F.C.; Lawal, H.A. Physicochemical analysis and mineral contents of honey from farmers in western states of Nigeria. *J. Nat. Sci. Res.* **2016**, *6*, 78–84.

34. Bosi, G.; Battaglini, M. Gas chromatographic analysis of free and protein amino acids in some unifloral honeys. *J. Apicult. Res.* **1978**, *17*, 152–166. [CrossRef]

35. Nozal, M.J.; Bernal, J.L.; Toribio, M.L.; Diego, J.C.; Ruiz, A. Rapid and sensitive method for determining free amino acids in honey by gas chromatography with flame ionization or mass spectrometric detection. *J. Chromatogr. A* **2004**, *1047*, 137–146. [CrossRef]

36. Bogdanov, S.; Baumann, S.E. Bestimmung von Honigzucker mit HPLC. *Mitt. Gebiete Lebensm. Hyg.* **1988**, *79*, 198–206.

37. Da Costa Leite, J.M.; Trugo, L.C.; Costa, L.S.M.; Quinteiro, L.M.C.; Barth, O.M.; Dutra, V.L.M.; De Maria, C.A.B. Determination of oligosaccharides in Brazilian honeys of different botanical origin. *Food Chem.* **2000**, *70*, 93–98. [CrossRef]

38. Rybak-Chmielewska, H.; Szczesna, T. Determination of saccharides in multifloral honey by means of HPLC. *J. Apicult. Sci.* **2003**, *47*, 93–101.

39. Ciulu, M.; Solinas, S.; Floris, I.; Panzanelli, A.; Pilo, M.I.; Piu, P.C.; Spano, N.; Sanna, G. RP-HPLC determination of water-soluble vitamins in honey. *Talanta* **2011**, *83*, 924–929. [CrossRef] [PubMed]

40. Edwards, R. A Department of Food Technology, University of New South Wales, Kensington, NSW, Australia.; Faraji-Haremi, R Department of Food Technology, University of New South Wales, Kensington, NSW, Australia.; Wootton, M Department of Food Technology, University of New South Wales, Kensington, NSW, Australia. A rapid method for determining the diastase activity in honey. *J. Apicult. Res.* **1975**, *14*, 47–50.

41. Hadorn, H. On the problem of the quantitative determination of diastase in honey. *Mitt. Gebiete Lebensm. Hyg.* **1961**, *52*, 67–103.

42. Bogdanov, S. Beeswax: Production, Properties, Composition and Control. Available online: http://www.bee-hexagon.net/files/file/fileE/Wax/WaxBook1.pdf (accessed on 25 September 2019).

43. Přidal, A. Beeswa–composition and utilization [in Czech]. *Moderní Včelař* **2007**, *5*, 20–21.

44. Mathivanan, V.; Shah, G.N.; Manzoor, M.; Mir, G.M. A review on propolis-as a novel folk medicine. *Indian J. Sci.* **2013**, *2*, 23–30.

45. Alban Muller International. *Phytami® Honey*; Raw material information data sheet; Alban Muller International: Chartres, France, 2009.

46. Crodarom. *Phytami® Honey*; Safety data sheet; Crodarom: East Yorkshire, UK, 2013.

47. CIOMS. *International Ethical Guidelines for Health-related Research Involving Humans*; Council for International Organizations of Medical Science (CIOMS): Geneva, Switzerland, 2016; Available online: https://cioms.ch/publications/product/international-ethical-guidelines-for-health-related-research-involving-humans/ (accessed on 20 September 2019).

48. International Organization for Standardization. *Sensory Analysis–Methodology–General Guidance*; ISO 6658; International Organization for Standardization: Geneva, Switzerland, 2017.

49. International Organization for Standardization. *Sensory Analysis General Guidance for the Design of Test Rooms*; ISO 8589; International Organization for Standardization: Geneva, Switzerland, 2017.

Sample Availability: Samples of the compounds are not available from the authors.

Review

Impact of Fermentation on the Phenolic Compounds and Antioxidant Activity of Whole Cereal Grains: A Mini Review

Oluwafemi Ayodeji Adebo [1,*] and **Ilce Gabriela Medina-Meza** [2]

1   Department of Biotechnology and Food Technology, Faculty of Science, University of Johannesburg, Doornfontein Campus, P.O. Box 17011, Gauteng, South Africa
2   Department of Biosystems and Agricultural Engineering, Michigan State University, 524 South Shaw Lane, East Lansing, MI 48824-1323, USA; ilce@msu.edu
*   Correspondence: oadebo@uj.ac.za; Tel.: +27-11-559-6261

Academic Editor: María Dolores Torres
Received: 22 December 2019; Accepted: 14 February 2020; Published: 19 February 2020

Abstract: Urbanization, emergence, and prominence of diseases and ailments have led to conscious and deliberate consumption of health beneficial foods. Whole grain (WG) cereals are one type of food with an array of nutritionally important and healthy constituents, including carotenoids, inulin, β-glucan, lignans, vitamin E-related compounds, tocols, phytosterols, and phenolic compounds, which are beneficial for human consumption. They not only provide nutrition, but also confer health promoting effects in food, such as anti-carcinogenic, anti-microbial, and antioxidant properties. Fermentation is a viable processing technique to transform whole grains in edible foods since it is an affordable, less complicated technique, which not only transforms whole grains but also increases nutrient bioavailability and positively alters the levels of health-promoting components (particularly antioxidants) in derived whole grain products. This review addresses the impact of fermentation on phenolic compounds and antioxidant activities with most available studies indicating an increase in these health beneficial constituents. Such increases are mostly due to breakdown of the cereal cell wall and subsequent activities of enzymes that lead to the liberation of bound phenolic compounds, which increase antioxidant activities. In addition to the improvement of these valuable constituents, increasing the consumption of fermented whole grain cereals would be vital for the world's ever-growing population. Concerted efforts and adequate strategic synergy between concerned stakeholders (researchers, food industry, and government/policy makers) are still required in this regard to encourage consumption and dispel negative presumptions about whole grain foods.

Keywords: fermentation; fermented foods; whole grains; health benefits; phenolic compounds; antioxidant activity

---

1. Introduction

Foods in the past were known to conventionally provide nutrients necessary for basic physiological functions. This assumption has changed with available knowledge at the disposal of consumers, changes in food regulations, and an ever-growing health-conscious population, which are factors resulting in an increasing desire for foods with additional physiological benefits. The 2500-year-old concept of "Let food be thy medicine and medicine be thy food" by Hippocrates is now being embraced better than ever as consumers are gradually becoming aware of the importance of diet in health promotion and disease prevention. Such a concept of food as medicine could have led to the trend of what is now known as "functional foods," which is a concept first created in Japan in the 1980s [1].

Supporting this perspective of food as medicine are several studies on whole grains (WGs) and WG-diets having positive effects on disease markers such as blood pressure, diabetes, and obesity [2–11].

WGs are essentially made up of the germ, bran, and endosperm and contains all the important parts of the entire grain seed in their original proportions. A more detailed and approved definition by the American Association of Cereal Chemists (AACC) says "WGs shall consist of the intact, ground, cracked, or flaked caryopsis, whose principal anatomical components—the starchy endosperm, germ, and bran—are present in the same relative proportions as they exist in the intact caryopsis" [12]. On the contrary, refined grains (RGs) are products obtained after the refining process involving the removal of the most potent protective components of the grains found in the bran and germ. This consequently leaves only the starchy-rich endosperm. The retained protective components in WGs make them better constituents of beneficial components as compared to their refined counterparts.

Health beneficial constituents of WGs include phytochemicals, bioactive carbohydrate fractions, peptides, and other phytonutrients [11,13–16]. WGs contain high amounts of phytochemicals, which are plant secondary metabolites that have shown biological activity and have been broadly investigated as health beneficial groups of compounds in food [17–19]. Particularly important are phenolic constituents, which are major forms of these phytochemicals and vital with reference to their unique contribution to the health benefits of WGs. The major sources of these phytochemicals are phenolic compounds (PCs) due to the high concentrations of bioactive constituents in the bran and germ layer [17,20,21] and the fact that they are largely one of the most important dietary sources of energy intake worldwide.

2. Phenolic Compounds in WG Foods

The overall benefit derived from three major components of WG (germ, bran, and endosperm) altogether is higher than any of the individual fractions [22,23]. A combination of these components makes WG contain physiologically important components including vitamins, fatty acids, phytosterols, PCs, fatty acids, dietary fiber, carotenoids, lignans, and sphingolipids (Figure 1), which can promote health either singly or in synergy with each other [18,24]. A series of meta analyses and multiple scientific studies have equally reported an association between increasing intake of WG-foods and reduced risk of non-communicable diseases such as cardiovascular diseases, coronary heart diseases, stroke [24–26], metabolic syndrome [27], and cancers [28,29] as well as a positive effect on gut microbiota [30]. Phenolic compounds are subsequently discussed in this review as it is of vital importance in WG-cereals [16] and the fact that they are the most studied phytochemicals [31]. Usually, WGs may be consumed as food after it has been incorporated as an ingredient into other food products or as food itself after processing. One type of such a food processing technique adopted for the transformation of WGs into diets is fermentation, which is a process that yields products that are not only shelf stable, but also better in sensorial qualities and health beneficial constituents [32–36]. The cereal bran is a major source of these PCs and this paper seeks to review available scientific literature on fermented WG-products to understand the influence and role of fermentation on PCs and antioxidant activity (AA) thereof.

Phenolic compounds (also called phenolics) are derived from several biosynthetic precursors including pyruvate, acetate, some amino acids (phenylalanine and tyrosine), malonyl CoA, acetyl CoA through the action of pentose phosphate, shikimate, and phenylpropanoid metabolism pathways [37–39]. The term 'phenolic acids' refers to phenolic compounds having one carboxylic acid group and are mainly divided into two subgroups, i.e., hydroxybenzoic acids (such as gallic, p-hydroxybenzoic, protocatechuic, syringic, and vanillic acids) and hydroxycinnamic acids (caffeic, ferulic, p-coumaric, and sinapic acids) (Figure 2). Flavonoids are an equally well-known class of frequently occurring phenolics in WGs. Major phenolics found in WGs are phenolic acids (PAs), flavonoids, and tannins. These plant-derived constituents are bioactive and involved in potentiating the redox defense of the body, prevention, and counteracting oxidative stress and reducing free radical-related cellular damage.

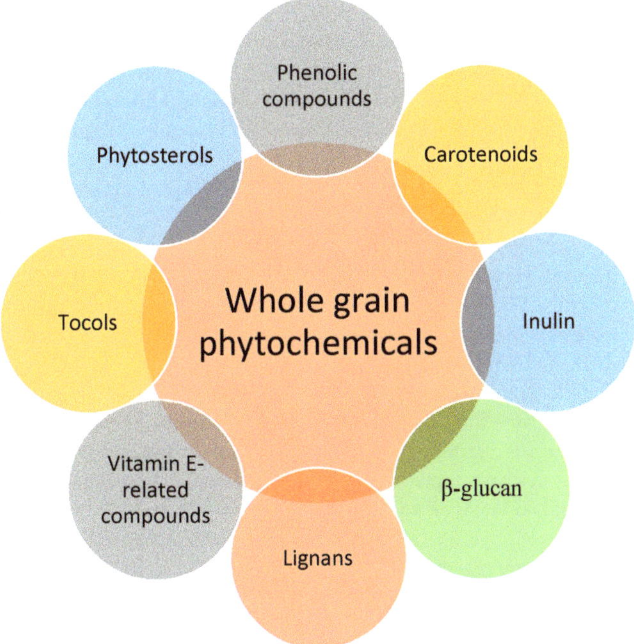

Figure 1. Whole grain phytochemicals.

As stated by Singh et al. [40], flavonoids are the largest group of phenolics and account for the half of known PCs in plants. These compounds are equally low molecular weight compounds consisting of two aromatic rings (A and B) joined by a three-carbon bridge (C_6–C_3–C_6 structure) [40]. Tannins, on the other hand, are high molecular weight polymeric phenolic compounds known to contribute to the pericarp (seed coat) color of cereals. These polyphenolic compounds have molecular weights of between 500–3000 g/mol, containing sufficient hydroxyls and other groups including carboxyl [41–43]. Tannins can be broadly classified into two, which include hydrolysable tannins [esters of ellagic acid (ellagitannins) or gallic acid (gallotannins)] and condensed tannins [(called polymeric proanthocyanidins) and known to be composed of flavonoid units) [41,44]. A plethora of excellent reviews and scientific literature are available in the literature on detailed classifications, forms, occurrences, and formation/generation of these compounds [15,16,40,41,45–50].

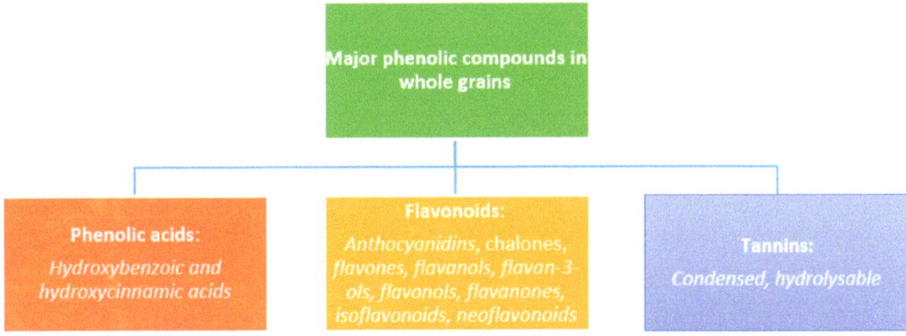

Figure 2. Classification of major phenolic compounds in whole grains.

3. Fermentation of WG Foods

Food processing is essential for the transformation of food crops into edible forms. Fermentation is an old food processing technique that has been adopted for centuries around the world, especially in developing nations. It involves an intentional conversion/modification of a substrate through activities of microorganisms to get a desired product. This is usually completed through microbial actions, which positively alter the appearance, flavor, functionalities, nutritional composition, color, and texture. The fermentation process itself yields beneficial effects through direct microbial action and production of metabolites and other complex compounds [51–53]. Conventional techniques of fermentation include (i) natural (also called spontaneous) occurrences through the actions of endogenous microorganisms, (ii) back slopping involves utilizing plenty of successful previous fermentation batches) and (iii) controlled fermentation, which entails the inoculation of starter cultures/specific strains. Subsequent fermented products are not only shelf stable through the preservative effect of this process, but fermentation also improves bioavailability and palatability, confers desirable organoleptic characteristics that impact aroma, texture, and flavor and improves the health beneficial components in food [32–36]. Irrespective of the food substrate (cereal, legume, vegetable, fruit, RG, or WG), fermentation results in the modification of inherent constituents, secondary metabolites, detoxification of toxic components/residues, and improvement in the functionality of the food product [35,36,53–55].

The incorporation of WG into diet which, is influenced by cultural beliefs, disadvantages of longer cooking time, the presence of phytates, tannins, and a limited variety of products made from them [56]. Additionally, some of their components may adversely affect the functional characteristics, taste, texture, and sensory appeal of subsequent formulations. Viable options for addressing this and incorporating WGs into diet would be completed through appropriate transformation into various other beneficial food forms, which would ensure the possibility of obtaining various value-added products. Although RGs are mostly used in fermented foods, the use of WGs as staple foods equally has a long history of human consumption [23]. Findings from epidemiological studies and discoveries, therefore, have triggered renewed interest among governmental bodies of different nations that WG should form part of cereal servings [24,57,58]. Table 1 summarizes common fermented WG products obtained through both solid-state fermentation (SSF) and liquid/submerged fermentation (SmF). While the former occurs in the absence or near-absence of free water, the latter occurs in the presence of free flowing water (more fluids compared to SSF). Subsequent fermented products are relatively few in contrast to numerous other studies reporting the use of RGs for similar food products, which necessitates further intensified research on the development of WG-fermented food products.

Table 1. Some reported fermented food products from whole grains.

Whole Grain(s)	Food	Type of Fermentation	Reference
Barley and oat	*Tempe*	SSF	Eklund-Jonsson et al. [59]
Maize	*Akamu/Ogi*	SSF	Oyarekua [60], Obinna-Echem et al. [61]
Millet	*Koji*	SSF	Salar et al. [62]
Millet	Probiotic drink	SmF	Di Stefano et al. [63]
Millet	Fermented milk	SmF	Sheela et al. [64]
Millet	Sourdough bread	SSF	Wang et al. [65]
Oat	Fermented oat	SSF	Wu et al. [66]
Oat, wheat	Bread	SSF	Gamel et al. [67]
Quinoa	Yoghurt	SmF	Zannini et al. [68]

Table 1. *Cont.*

Whole Grain(s)	Food	Type of Fermentation	Reference
Quinoa, wheat	Fermented product	SSF	Ayyash et al. [69]
Rye, oat, wheat	Bread	SSF	Buddrick et al. [70]
Rye, wheat	Sourdough bread	SSF	Koistinen et al. [71]
Rye	Bread	SSF	Johansson et al. [72], Raninen et al. [73]
Rye	Porridge	SSF	Lee et al. [74]
Rye	Sourdough bread	SSF	Beckmann et al. [75], Zamaratskaia et al. [76]
Sorghum	*Burukutu*	SmF	Ikediobi et al. [77]
Sorghum	Fermented balls	SSF	Ragaee and Abdel-Aal [78]
Sorghum	Fermented porridge	SSF	Dlamini et al. [79]
Sorghum	*Injera*	SSF	Taylor and Taylor [80]
Sorghum	*Ogi*	SmF	Akingbala et al. [81]
Sorghum	*Omuramba*	SmF	Mukuru et al. [82]
Sorghum	*Ting*	SSF	Kruger et al. [83], Adebo et al. [84,85]
Sorghum	*Uji*	SmF	Taylor and Taylor [80]
Tef	*Injera*	SSF	Tamene et al. [86]
Wheat	*Boza*	SmF	Gotcheva et al. [87]
Wheat	Bread	SSF	Mustafa and Adem [88], Struyf et al. [89]
Wheat	Sourdough bread	SSF	García-Mantrana et al. [90]
Wheat	*Tempe*	SSF	Dey and Kuhad [91], Starzyńska-Janiszewska et al. [92]

SSF—solid-state fermentation. SmF—submerged/liquid fermentation.

Due to the protective pericarp/seed coat, the fermentation process might be slightly hindered. Such has been reported in the literature and attributed to some of the antimicrobials and bioactive constituents in the seed coat that might mitigate the activity of fermenting microorganisms [55,90,93,94]. The protective pericarp layer of cereal tends to alter the diffusion of nutrients such as amino acids and sugars necessary for the growth of fermenting microorganisms. While this might result in a slightly higher pH and likely longer fermentation periods (in the absence of a starter culture), fermentation still modifies the phenolic constituents in WGs.

4. Impact of Fermentation on Phenolic Compounds in WGs

The fermentation process can have multiple effects on WG phenolics leading to modifications in inherent levels and/or formation of subsequent monomers or polymers. Adebo et al. [84] reported higher bioactive compounds (catechin, gallic acid, and quercetin) after fermentation in a study on ting from fermented WG-sorghum with a concurrent decrease in total flavonoid content (TFC), total tannin content (TNC), and total phenolic content (TPC). Reported decreases in levels of TPC, TFC, and TNC were attributed to degradation and hydrolysis of the phenolic compounds, while a corresponding increase in catechin, gallic acid, and quercetin was attributed to a release of these bioactive compounds after fermentation with *Lactobacillus* strains.

Through fungal fermentation of WG-wheat into *tempe*, an increase in the sum of PAs was observed with up to a 382% increase in ferulic acid recorded after fermentation [92]. A similar trend of increase in investigated PCs and TPC during the fermentation of WG-*tempe* with *Rhizopus oryzae* RCK2012 had been reported earlier [91]. Salar et al. [62] equally reported an increase in TPC of the WG-millet-*koji* and attributed this to mobilization of PCs from their bound form to a free state through enzymes produced during fermentation. Similar authors earlier reported an increase in TPC during the fermentation of WG-maize [95], reportedly through the activities of β-glucosidase, which is capable of hydrolyzing phenolic phucosides to release free phenolics. Increased extractability of PCs, synthesis of new bioactive compounds, and consequent liberation of PCs due to structural breakdown of cereal cell walls have all been attributed to such increases in WG-PCs after fermentation (Table 2). Through metabolic activities of microbes, fermentation also induces structural breakdown of the cell wall, which leads to synthesis of various bioactive compounds [65]. Equally important are the roles of proteases, amylases, xylanases derived from fermenting microorganisms, and the cereal grain that contributes to modification of the grain and distorting of chemical bonds, which, consequently, releases bound phenolics (Figure 3).

Table 2. Documented studies on the effect of fermentation on phenolics of whole grains.

Whole Grain	Fermented Product	Phenolics Investigated	Analytical Method	Findings	References
Barley and oat groat	Fermented product	Free and bound PAs	Colorimetric; HPLC	Increase in total content of bound PAs in flours from WG-barley related to increased content of bound *p*-coumaric acid, ferulic acid, and dimers of ferulic acid (5,5'- diferulic, 8-*o*-4-diferulic, and 8,5'-diferulic acids).	Hole et al. [96]
Maize	Fermented product	TPC	Colorimetric	Increase in TPC after fermentation attributed to the activities of β-glucosidase, capable of hydrolyzing phenolic phucosides to release free phenolics	Salar et al. [95]
Millet	*Koji*	TPC	Colorimetric	Increase in TPC of fermented product due to mobilization of PCs from their bound form to a free state through enzymes produced during fermentation	Salar et al. [62]
Millet	Sourdough bread	TPC	Colorimetric	Increase and decrease in soluble and bound phenolic content. Slight decrease in TPC observed. Increment of soluble phenolic content may be due to acidification, production of hydrolytic enzymes by LAB, and/or activation of indigenous cereal enzymes, which broke down the bran cell wall structure	Wang et al. [65]
Quinoa, wheat	Fermented product	TPC	Colorimetric	Increase in TPC may be attributed to hydrolytic activities (e.g., esterases) of Bifidobacteria strains that released more PCs via the hydrolysis of complexed forms, possibly the synthesis of new bioactive compounds detected as PCs	Ayyash et al. [69]
Rye	Baked sourdough	TPC, PAs	Colorimetric, HPLC	Fermentation phase more than doubled the levels of easily extractable PCs	Liukkonen et al. [97]
Rye	Sourdough	TPC, PAs	Colorimetric, HPLC	Increased level of total PCs due to increases in methanol-extractable PCs. Modification in levels of bioactive compounds during fermentation by the metabolic activity of microbes. Fermentation-induced structural breakdown of cereal cell walls might have also occurred and led to liberation and/or synthesis of various bioactive compounds	Katina et al. [98]

Table 2. *Cont.*

Whole Grain	Fermented Product	Phenolics Investigated	Analytical Method	Findings	References
Rye, wheat	Whole meal bread	PAs	HPLC	Increase in PAs due to activities of phenolic acid esterases during the fermentation stage	Skrajda-Brdak et al. [99]
Sorghum	Fermented porridge	TPC, TNC	Colorimetric	Reduction in TNC and TPC. Reduction in TNC could be due to binding of tannins with protein and other components, which reduces their extractability and tannin degradation by microbial enzymes	Dlamini et al. [79]
Sorghum	Fermented product	TPC, TNC	Colorimetric	Increase in TPC, decrease in TNC	Mohapatra et al. [100]
Sorghum	*Ting*	Flavonoids, PA, TFC, TNC, TPC	Colorimetric, LC-MS/MS	Decrease in TFC, TNC, and TPC attributed to possible degradation of PCs and hydrolysis of bioactive compounds. Breakdown of tannin-related compounds to lower molecular weight compounds, which affected extractability. Increase in PA and flavonoids could be due to decarboxylation, hydrolysis, microbial oxidation, and reduction as well as esterification reactions that occurred during fermentation	Adebo et al. [84,85]
Wheat	Fermented product	TPC	Colorimetric	Increase in TPC through modification in levels of bioactive compounds during fermentation by the metabolic activity of microbes	Đordevic et al. [101]
Wheat	Sourdough	PAs	LC-MS/MS, UPLC	Degradation, reduction of some PAs and content of some remain unchanged. Release of PAs from bound fraction, metabolism of PA by LAB strains and action of enzymes (decarboxylases, esterases, and reductases)	Ripari et al. [102]
Wheat	*Tempe*	TPC, PCs	Colorimetric, TLC, UPLC	Increase in TPC after fermentation, possibly due to release of bound compounds from the wheat matrix	Dey and Kuhad [91]
Wheat	*Tempe*	Free and condensed PAs	HPLC	Increase in the sum of PA could be linked to an increase in their extractability after fermentation	Starzyńska-Janiszewska et al. [92]
Wheat, brown rice, maize, oat	Fermented product	TPC, PAs	Colorimetric, HPLC	TPC of all fermented samples increased except for *Rhizopus oligosporus* fermented maize. Increase as well as decrease in PA levels. Decreases was attributed to strain/specie specificity and/or grain composition. General increases were alluded to enhanced bioavailability of cereal phenolics.	Dey and Kuhad [103]

HPLC—high performance liquid chromatography. LAB—lactic acid bacteria. LC-MS/MS – liquid chromatography tandem mass spectrometry. PA—phenolic acid. PC—phenolic compound. TFC—total flavonoid content. TLC—thin layer chromatography. TNC—total tannin content. TPC—total phenolic content. UPLC—ultra high-performance liquid chromatography.

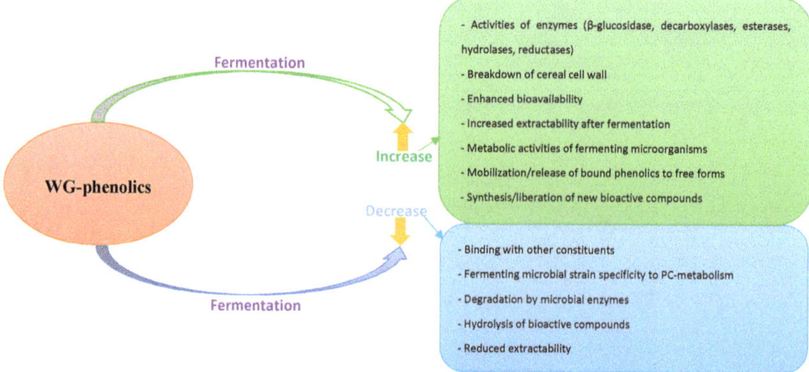

Figure 3. A summary of ways by which whole grain phenolic compounds are modified during fermentation.

During fermentation, PCs are metabolized and modified by fermenting organisms into other conjugates, glucosides, and/or related forms. Such a metabolism of PCs during fermentation have been reported to increase their bioavailability [104,105] and lead to generation of compounds that impact flavor [106,107]. Fermentation of sorghum into sourdough using LAB strains [singly and in two binary combinations (*L. plantarum* and *L. casei* or *L. fermentum* and *L. reuteri*)] was reported to have resulted in the metabolism of PAs, PA-esters, and flavonoid glucosides [108]. Most PCs in this study were metabolized and most notable were the transformation of caffeic acid → dihydrocaffeic acid, ethylcatechol, vinylcathechol, ferulic acid → dihydroferulic acid and naringenin-7-*O*-glucoside → naringenin, reportedly an indication of the presence of esterase (tannase), glucosidase, PA decarboxylase, and PA reductase [108]. The authors also suggested that the strains might have used different pathways for PA and flavonoid metabolism. Fermentation of WG-sorghum have also been reported to have led to the modification of PCs (catechin, gallic acid, and quercetin) into structurally related compounds, which were not identified [85]. The authors suggested that the observed modification could be attributed to decarboxylation, hydrolysis, and esterification reactions that might have occurred during fermentation [85]. In a study on the metabolism of PAs in whole wheat and rye malt sourdoughs, *L. plantarum* was observed to have metabolized free ferulic acid in wheat and rye malt sourdoughs, while a strain of *L. hammesii* (DSM 16381) metabolized syringic and vanillic acids and reduced levels of bound ferulic acid in wheat sourdoughs [102]. Co-fermentation of the LAB strains was also noted to have aided the conversion of resultant-free ferulic acid to dihydroferulic acid and volatile metabolites (vinyl-guaiacol and ethyl-guaiacol), which suggests that PA metabolism in sourdoughs is more enhanced by co-fermentation due to complementary metabolic activities [102]. Carboxylase, decarboxylase, esterase, and reductase activities in the LABs were reportedly responsible for PA metabolism in this study [102]. It should, however, be noted that such metabolism could lead to an increase in antimicrobial activities of resulting metabolic products [109], a decrease in antimicrobial activities [104,110], or no alteration in antimicrobial activity levels [108].

According to Gänzle [104], metabolism of PCs may involve the removal of noxious compounds as well as the release of hexosides as a source of metabolic energy. This metabolism could, however, be influenced by composition and intrinsic factors of the matrices/substrate and can, thus, influence the metabolic pathway, i.e., enzymatic activities can shift from decarboxylase action to reductase to glucosidase activity [111]. Glycosyl hydrolases have also been implicated as a group of enzymes responsible for such metabolism of PCs [104]. For example, *L. hammesii* was reported to have metabolized hydroxybenzoic acids in wheat but not in rye malt sourdoughs, which possibly reflects that the fermentation substrate influences the expression of enzymes active on PAs [111]. Likewise, in a study on sorghum sourdough, the accumulation of dihydrocaffeic acid by only *L. fermentum* indicates that decarboxylase and reductase enzymes of the other strains (*L. fermentum* and *L. plantarum*) have different substrate specificities [108]. The study of Gaur et al. [112] also suggests that availability of genes necessary for the metabolism of these PCs is also of importance and a significant contributor to the metabolic potential of fermenting microorganisms.

5. Impact of Fermentation on Antioxidant Activity in WGs

Antioxidants are endogenous or exogenous molecules that mitigate any form of oxidative/nitrosative stress or its consequences [113]. According to Slavin [114], the primary protective role of antioxidants in the body is through their reaction with free radicals. Antioxidants function as free radical scavengers, quenchers of singlet oxygen formation, and reducing agents [115,116] through their inhibitory activity of prooxidant enzymes. A potential mechanism by which PCs confer AA involves the induction of detoxification mechanisms through phase II conjugation reactions, which prevents the formation of carcinogens from precursors as well as by blocking the reaction of carcinogens with critical cellular macromolecules [117,118]. Phenolic compounds also modify some cellular signaling processes and donate an electron/transfer hydrogen atom to free radicals, activate endogenous antioxidant mechanisms, which increases the levels of antioxidant enzymes, and act as chelators of trace metals involved in free radical protection [116,119,120].

As evident in Table 3, most available studies in the literature investigating the influence of fermentation on phenolic compounds have majorly focused on AAs as its health benefit. This might be unsurprising as PCs, particularly PAs, have been reported as one of the most abundant metabolites of cereal crops with AAs [121–123]. While the role of other bioactive constituents in WGs cannot be disregarded, PCs equally play a huge role in the antioxidant properties it confers to WG-foods.

Table 3. Documented studies on the effect of fermentation on antioxidant activity of whole grains.

Whole Grain	Fermented Product	Assay	Mechanism(s) Reported	References
Maize	Fermented product	ABTS, DPPH	Increase in ABTS and DPPH due to the role of the hydrolytic enzyme that released/mobilized bound polyphenolic compounds, which enhanced AAs.	Salar et al. [95]
Millet	*Koji*	ABTS, DPPH	*Koji* showed increased scavenging of ABTS and DPPH radicals due to the release of a bound form of phytochemicals present and high levels of TPC modulated during fermentation.	Salar et al. [62]
Millet	Sourdough bread	DPPH	Increase in DPPH radical inhibition after sourdough fermentation. The conversion of bound to soluble PCs improved the health-related functionality of the final products.	Wang et al. [65]
Quinoa, wheat	Fermented product	ABTS, DPPH	An increase in ABTS and DPPH values was attributed to the soluble phytochemicals released during fermentation and to bioactive peptides formed as a result of proteolytic activity.	Ayyash et al. [69]
Rye	Baked sourdough	DPPH	The fermentation stage increased AA likely due to an increased level of extractable PCs.	Liukkonen et al. [97]

Table 3. *Cont.*

Whole Grain	Fermented Product	Assay	Mechanism(s) Reported	References
Sorghum	Fermented porridge	ABTS, DPPH	Reduction in antioxidant levels after fermentation attributed to changes during processing that affected the extraction of total phenols and tannins. Such changes were hypothesized to have likely involved associations between the tannins, phenols, proteins, and other compounds in the grain.	Dlamini et al. [79]
Sorghum	Fermented product	CUPRAC, DPPH	Increase in AAs investigated.	Mohapatra et al. [100]
Sorghum	*Ting*	ABTS	Increase in AA due to regenerated and released bioactive compounds (including non-phenolic components after fermentation with the *L. fermentum* strains), which might have contributed to the radical scavenging properties of the product.	Adebo et al. [85]
Wheat	Fermented product	DPPH, FRAP, TBA	Increase in the investigated AAs.	Đordevic et al. [101]
Wheat	*Tempe*	ABTS, DPPH, FRAP, HP-scavenging and OH-scavenging assays	Increase in antioxidant properties investigated attributed to the composition of PCs, unidentified compounds, and other water-soluble bioactive compounds like small peptides and xylo-oligosaccharides produced during fermentation.	Dey and Kuhad [91]
Wheat	*Tempe*	ABTS, OH-scavenging and FCRS-RP assays	Increase in soluble antioxidant potential as fermentation increased extractable antiradical activity scavenging potential, which might be due to the release of peptides and other compounds during fermentation.	Starzyńska-Janiszewska et al. [92]
Wheat, brown rice, maize, oat	Fermented product	ABTS, DPPH	Both ABTS and DPPH scavenging properties were enhanced after fermentation of the WG-cereals by all the four micro-organisms (except *R. oligosporus*-fermented maize). Increases related to release of more soluble bioactive compounds, such as peptides and oligosaccharides.	Dey and Kuhad [103]

ABTS-2,2'-azino-bis(3-ethylbenzothiazoline-6-sulfonic acid). CUPRAC—cupric reducing antioxidant capacity. DPPH—2,2-diphenyl-1-picrylhydrazyl. FCRS-RP—Folin-Ciocalteu reacting substances-reducing power. FRAP—ferric reducing antioxidant property. HP—hydrogen peroxide. HPLC—high performance liquid chromatography. OH—hydroxyl.

Although the majority of the studies reviewed herein reported increases in PCs, this is not always the case, as decreases in these health beneficial constituents have also been reported (Table 2). Studies on fermented WG-sorghum reported a decrease in TNC and TPC with this attributed to the ability of tannins to bind with proteins and other components, which reduces extractability as well as tannin degradation [79,85]. Investigations into the metabolism of sourdough by Ripari et al. [102] also suggested that reduction in some investigated PAs might be due to metabolism of PAs by lactic acid bacteria (LAB) and the activities of decarboxylases, esterases, and reductases. In the study of Dey and Kuhad [103] on fermentation of different WGs, both an increase and a decrease in TPC was observed. While increases alluded to enhanced bioavailability of cereal phenolics, a decrease observed in maize was associated with the specificity of the microbial strain to act on the PCs as well as the grain composition. The effect of the microbial activity on the levels of individual phenolics can differ, depending on the microbial strain. The genome of certain microorganisms might encode genes responsible for the metabolism and/or degradation of phenolic compounds while some do not [92,96,102]. This might, however, be difficult to ascertain or distinguish in spontaneous fermentation processes or back-slopping that is characterized by a wide range of fermenting microorganisms.

During the estimation of AA of food products, using more than one analytical method is better because food contains a myriad of constituents [92]. The frequently used techniques are spectrophotometric assays and the 2, 2′-Azino-bis (3-ethylbenzothiazoline-6-sulfate) (ABTS) (also called ABTS-radical cation depolarization) assay as well as the cupric-reducing antioxidant capacity (CUPRAC), 2,2-Diphenyl-1-picrylhydrazyl (DPPH) and ferric-reducing antioxidant power (FRAP) assay. Less frequently used techniques found in the course of this review are the lipid peroxidation technique adopting the thiobarbituric acid (TBA) assay, which was used to determine the TBA reactive substance from lipid peroxidation [101], as well as OH- and H_2O_2-scavenging assays. These are both concerning due to their role in causing tissue damage and cell death, and could combine with nucleotides to cause carcinogenesis [124].

Considering the general trend of increase in WG-PCs after fermentation and associated mechanisms, it could, thus, be hypothesized that this should be tantamount to an increase in AAs. While such increases were reported, some studies noted decreases in AAs of WG-fermented products. As documented by Đordevic et al. [101] and Sun and Ho [125], possible explanations for this ambiguous relationship between AA and PCs are that: (i) quantified TPC values do not include other components that can equally confer AAs, (ii) synergy in a mixture makes AA not only dependent on antioxidant concentration but also on the structure and interactions among antioxidants, and (iii) different methods used for measuring AA based on different mechanisms may lead to different observations. Such an observation has also been buttressed by other authors suggesting that directly linking AAs in food and a responsible component might be somewhat difficult, as methods of extraction, identification, and/or quantification of AAs vary [126,127], which makes comparisons and, subsequently, extrapolating conclusions quite tricky.

General increases in AA of fermented foods have been attributed to a release of bound PC due to activities of hydrolytic enzymes and contents modulated during fermentation of a maize-based product and *koji* from millet [62,95]. A likely conversion of bound PCs into health-related components, a release of soluble phytochemicals and other non-PCs as well as increased extractability of AA-related PCs have equally been implicated to have led to an increase in AA during the fermentation of WGs into tempe, ting, and sourdough (from millet and rye) [65,69,85,92,97]. An addition to these could be that the fermentation process facilitated cleavage/dissociation of the bonds between PCs and other constituents leading to a release of PC-monomers, which yield AAs. Equally important and implicated in other studies are products of protein hydrolysis through proteolytic actions through fermentation, which could have led to components that contribute to increased PC and consequent antioxidant potential of fermented WGs. Available enzymes during fermentation and/or produced by fermenting microorganisms could also break down ester bonds, hydrolyse β-glucosidic bonds, and distort the hydroxyl groups in phenolic structures liberating free PCs and other antioxidant-related compounds. On the contrary, a decrease in AA after fermentation was attributed to modifications that influenced the extractability of compounds that confer AAs, especially the association between tannins, phenols, proteins, and other compounds in the grain [79].

Although in vitro studies reflect potential AAs of WG-fermented cereals, these in vitro techniques could underestimate physiological antioxidants, which necessitates in vivo studies. The use of in vivo models in investigating the influence of fermentation on AA is largely desirable. According to Benedetti et al. [128] and Alam et al. [129], in vivo protocols involve the administration of antioxidants to testing animals for a specified period of time, after which the animals are sacrificed, and blood or tissues are analyzed. Subsequently done are assays such as lipid peroxidation (LPO), thioredoxin reductase activities, and glutathione peroxidase (GSHPx) in human patients [128,130]. Although such in vivo studies are largely desirable, challenges related to ethical approvals, high costs, and daunting logistics have led to the adoption of in vitro techniques. Few studies are available on in vivo assays on fermented WG-cereal products with such studies focusing on AAs of the product. Breads made from WG-Kamut Khorasan wheat and WG-durum wheat were both reported to protect rat liver from oxidative stress [128]. An earlier study by similar authors reported a lower oxidative state in rats fed with experimental diets of sourdough bread for seven weeks [131].

Phenolic compounds usually occur in an esterified form linked to the cell wall matrix in the cereal bran and, as such, not readily available. Fermentation is considered a possible strategy to not only increase AAs but also to release the insoluble bound phenolic acids and, thus, to improve the poor bioavailability of grain phenolics [132]. This is particularly important as the antioxidant potential of WGs could be restricted by low availability of compounds during digestion. Not only does fermentation increases PCs and AA of WG-fermented products (Tables 2 and 3), it also positively influences bioavailability, bio-accessibility, and PAs as demonstrated in a study on flours from WG-barley fermented with probiotic strains [96].

6. Future Perspective

Fermentation positively alters food quality, confers organoleptic characteristics, and improves phenolic constituents and antioxidant activity of WGs. Could this then translate to consumption of more whole grains? Possibly not, considering the grittiness and associated sensory challenges associated with whole grain foods. This might also contribute to fewer whole grain fermented foods as compared to those from refined grains. This is in tandem with a study on the consumption of WGs foods from brewers' spent grain, which indicates that hereditary consumers of whole grain foods will be more receptive to its consumption as compared to their refined foods counterpart [133]. Some studies have also indicated barriers for consuming WG foods such as the lack of knowledge about its health benefits, challenges with cooking/preparation time, negative sensory perception, perceived cost, and the lack of availability of whole grains [134–136].

7. Conclusions

Increasing whole grain consumption should, therefore, be a target for health organizations with recommendations for intake proposed in many countries. As such, new strategies and partnerships between researchers, industry, and relevant agencies are further needed to promote whole grain consumption. Future studies are necessary in the area of phenolic compounds in fermented whole grains along with effective techniques such as whole genome sequencing to investigate genes responsible for the conversion of phenolic constituents and improvements in AAs. Such would largely assist in choosing starter cultures that would further improve the quality of fermented WG foods. Deeper investigation into the mechanisms of different forms of fermentation (solid state and liquid) on single/pure phenolic compounds (in isolation) and antioxidant activities should equally be explored. Additionally, studies are needed into the absorption and bioavailability of these phenolics in the gut, preferably through in vivo models.

Funding: The University of Johannesburg Global Excellence and Stature (GES) 4.0 Catalytic Initiative Grant and the National Research Foundation (NRF) of South Africa Thuthuka funding (Grant no: 121826) are duly acknowledged.

Conflicts of Interest: The authors declare no conflict of interest.

References

1. Arai, S. Studies on functional foods in Japan-state of the art. *Biosci. Biotechnol. Biochem.* **1996**, *60*, 9–15. [CrossRef] [PubMed]
2. Hallfrisch, J.; Scholfield, D.J.; Behall, K.M. Blood pressure reduced by whole grain diet containing barley or whole wheat and brown rice in moderately hypercholesterolemic men. *Nutr. Res.* **2003**, *23*, 1631–1642. [CrossRef]
3. Juntunen, K.S.; Laaksonen, D.E.; Poutanen, K.S.; Niskanen, L.K.; Mykkanen, H.M. High-fiber rye bread and insulin secretion and sensitivity in healthy postmenopausal women. *Am. J. Clin. Nutr.* **2003**, *77*, 385–391. [CrossRef] [PubMed]

4. Behall, K.M.; Scholfield, D.J.; Hallfrisch, J. Diets containing barley significantly reduce lipids in mildly hypercholesterolemic men and women. *Am. J. Clin. Nutr.* **2004**, *80*, 1185–1193. [CrossRef] [PubMed]

5. Behall, K.M.; Scholfield, D.J.; Hallfrisch, J. Whole-grain diets reduce blood pressure in mildly hypercholesterolemic men and women. *J. Am. Diet. Assoc.* **2006**, *106*, 1445–1449. [CrossRef] [PubMed]

6. Marquart, L.; Jones, J.M.; Chohen, E.A.; Poutanen, K. The future of whole grains. In *Whole Grains and Health*; Marquart, L., Jacobs, D.R., McIntosh, G.H., Poutanen, K., Reicks, M., Eds.; Blackwell Publishing: Ames, IA, USA, 2007; pp. 3–16.

7. Priebe, M.G.; van Binsbergen, J.J.; de Vos, R.; Vonk, R.J. Whole grain foods for the prevention of type 2 diabetes mellitus. *Cochrane Database Syst. Rev.* **2008**, *23*, CD006061. [CrossRef]

8. Maki, K.C.; Beiseigel, J.M.; Jonnalagadda, S.S.; Gugger, C.K.; Reeves, M.S.; Farmer, M.V.; Kaden, V.N.; Rains, T.M. Whole-grain ready-to-eat oat cereal, as part of a dietary program for weight loss, reduces low-density lipoprotein cholesterol in adults with overweight and obesity more than a dietary program including low-fiber control foods. *J. Am. Diet. Assoc.* **2010**, *110*, 205–214. [CrossRef]

9. Tighe, P.; Duthie, G.; Vaughan, N.; Brittenden, J.; Simpson, W.G.; Duthie, S.; Mutch, W.; Wahle, K.; Horgan, G.; Thies, F. Effect of increased consumption of whole-grain foods on blood pressure and other cardiovascular risk markers in healthy middle-aged persons: A randomized controlled trial. *Am. J. Clin. Nutr.* **2010**, *92*, 733–740. [CrossRef]

10. Ross, A.B.; Bruce, S.J.; Blondel-Lubrano, A.; Oguey-Araymon, S.; Beaumont, M.; Bourgeois, A.; Nielsen-Moennoz, C.; Vigo, M.; Fay, L.B.; Kochhar, S.; et al. A whole-grain cereal-rich diet increases plasma betaine and tends to decrease total and LDL-cholesterol compared with a refined-grain diet in healthy subjects. *Br. J. Nutr.* **2011**, *105*, 1492–1502. [CrossRef]

11. Seal, C.J.; Brownlee, I.A. Whole-grain foods and chronic disease: Evidence from epidemiological and intervention studies. *Proc. Nutr. Soc.* **2015**, *74*, 313–319. [CrossRef]

12. AACC. Definition of Whole Grain. Available online: https://www.cerealsgrains.org/initiatives/definitions/Pages/WholeGrain.aspx (accessed on 8 November 2019).

13. Jonnalagadda, S.S.; Harnack, L.; Liu, R.H.; McKeown, N.; Seal, C.; Liu, S.; Fahey, G.C. Putting the whole grain puzzle together: Health benefits associated with whole grains-Summary of American Society for Nutrition 2010 satellite symposium. *J. Nutr.* **2011**, *141*, 1011S–1022S. [CrossRef] [PubMed]

14. Borneo, R.; León, A.E. Whole grain cereals: Functional components and health benefits. *Food Funct.* **2012**, *3*, 110–119. [CrossRef] [PubMed]

15. McRae, M.P. Health benefits of dietary whole grains: An umbrella review of meta-analyses. *J. Chiroprac. Med.* **2017**, *16*, 10–18. [CrossRef] [PubMed]

16. Călinoiu, L.F.; Vodnar, D.C. Whole grains and phenolic acids: A review on bioactivity, functionality, health benefits and bioavailability. *Nutrients* **2018**, *10*, 1615. [CrossRef] [PubMed]

17. Liu, R.H. Whole grain phytochemicals and health. *J. Cereal. Sci.* **2007**, *46*, 207–219. [CrossRef]

18. Okarter, N.; Liu, R.H. Health benefits of whole grain phytochemicals. *Crit. Rev. Food Sci. Nutr.* **2011**, *50*, 193–208. [CrossRef]

19. Zhu, Y.; Sang, S. Phytochemicals in whole grain wheat and their health-promoting effects. *Mol. Nutr. Food Res.* **2017**, *61*, 1600852. [CrossRef]

20. Koistinen, V.M.; Katina, K.; Nordlund, E.; Poutanen, K.; Hanhineva, K. Changes in the phytochemical profile of rye bran induced by enzymatic bioprocessing and sourdough fermentation. *Food Res. Int.* **2016**, *89*, 1106–1115. [CrossRef]

21. Idehen, E.; Tang, Y.; Sang, S. Bioactive phytochemicals in barley. *J. Food Drug Anal.* **2017**, *25*, 148–161. [CrossRef]

22. Harvard Health Letter. The whole is greater than the sum of the parts. I. Whole grain. *Harv. Health Lett.* **1999**, *25*, 8.

23. Schaffer-Lequart, C.; Lehmann, U.; Ross, A.B.; Roger, O.; Eldridge, A.L.; Ananta, E.; Bietry, M.F.; King, L.R.; Moroni, A.V.; Srichuwong, S.; et al. Whole grain in manufactured foods: Current use, challenges and the way forward. *Crit. Rev. Food Sci. Nutr.* **2017**, *57*, 1562–1568. [CrossRef] [PubMed]

24. Jones, J.M.; Reicks, M.; Adams, J.; Fulcher, G.; Weaver, G.; Kanter, M.; Marquart, L. The importance of promoting a whole grain foods message. *J. Am. Coll. Nutr.* **2002**, *21*, 293–297. [CrossRef] [PubMed]

25. Mellen, P.B.; Walsh, T.F.; Herrington, D.M. Whole grain intake and cardiovascular disease: A meta-analysis. *Nutr. Metab. Cardiovasc. Dis.* **2008**, *18*, 283–290. [CrossRef] [PubMed]

26. Ye, E.Q.; Chacko, S.A.; Chou, E.L.; Kugizaki, M.; Liu, S. Greater whole-grain intake is associated with lower risk of type 2 diabetes, cardiovascular disease, and weight gain. *J. Nutr.* **2012**, *142*, 1304–1313. [CrossRef] [PubMed]

27. Esmaillzadeh, A.; Mirmiran, P.; Azizi, F. Whole-grain consumption and the metabolic syndrome: A favorable association in Tehranian adults. *Eur. J. Clin. Nutr.* **2005**, *59*, 353–362. [CrossRef] [PubMed]

28. Jacobs, D.; Marquart, L.; Slavin, J.; Kushi, L.H. Whole grain intake and cancer: An expanded review and meta-analysis. *Nutr. Cancer* **1998**, *30*, 85–96. [CrossRef]

29. Mourouti, N.; Papavagelis, C.; Plytzanopoulou, P.; Kontogianni, M.; Vassilakou, T.; Malamos, N.; Linos, A.; Panagiotakos, D. Dietary patterns and breast cancer: A case-control study in women. *Eur. J. Nutr.* **2015**, *54*, 609–617. [CrossRef]

30. Vanegas, S.M.; Maydani, M.; Barnett, J.B.; Goldin, B.; Kane, A.; Rasmussen, H.; Brown, C.; Vangay, P.; Knights, D.; Jonnalagadda, S.; et al. Substituting whole grains for refined grains in a 6-wk randomized trial has a modest effect on gut microbiota and immune and inflammatory markers of healthy adults. *Am. J. Clin Nutr.* **2017**, *105*, 635–650. [CrossRef]

31. Boudet, A.M. Evolution and current status of research in phenolic compounds. *Phytochemistry* **2007**, *68*, 2722–2735. [CrossRef]

32. Arendt, E.K.; Moroni, A.; Zannini, E. Medical nutrition therapy: Use of sourdough lactic acid bacteria as a cell factory for delivering functional biomolecules and food ingredients in gluten free bread. *Microb. Cell Fact.* **2011**, *10*, S15. [CrossRef]

33. Coda, R.; Cagno, R.D.; Gobbetti, M.; Rizzello, C.G. Sourdough lactic acid bacteria: Exploration of non-wheat cereal-base fermentation. *Food Microbiol.* **2014**, *37*, 51–58. [CrossRef]

34. Adebo, O.A.; Njobeh, P.B.; Adeboye, A.S.; Adebiyi, J.A.; Sobowale, S.S.; Ogundele, O.M.; Kayitesi, E. Advances in fermentation technology for novel food products. In *Innovations in Technologies for Fermented Food and Beverage Industries*; Panda, S.K., Shetty, P.H., Eds.; Springer: Cham, Switzerland, 2018; pp. 71–87.

35. Adebo, O.A.; Kayitesi, E.; Tugizimana, F.; Njobeh, P.B. Differential metabolic signatures in naturally and lactic acid bacteria (LAB) fermented *ting* (a Southern African food) with different tannin content, as revealed by gas chromatography mass spectrometry (GC–MS)-based metabolomics. *Food Res. Int.* **2019**, *121*, 326–335. [CrossRef]

36. Adebiyi, J.A.; Kayitesi, E.; Adebo, O.A.; Changwa, R.; Njobeh, P.B. Food fermentation and mycotoxin detoxification: An African perspective. *Food Cont.* **2019**, *106*, 106731. [CrossRef]

37. Robards, K.; Prenzler, P.D.; Tucker, G.; Swatsitang, P.; Glover, W. Phenolic compounds and their role in oxidative process in fruits. *Food Chem.* **1999**, *66*, 401–436. [CrossRef]

38. Shadidi, F. Phytochemicals in oilseeds. In *Phytochemicals in Nutrition and Health*; CRC Press: Boca Raton, FL, USA, 2002; pp. 139–156.

39. Randhir, R.; Lin, Y.-T.; Shetty, K. Phenolics, their antioxidant and antimicrobial activity in dark germinated fenugreek sprouts in response to peptide and phytochemical elicitors. *Asia Pac. J. Clin. Nutr.* **2004**, *13*, 295–307.

40. Singh, B.; Singh, J.P.; Kaur, A.; Singh, N. Phenolic composition and antioxidant potential of grain legume seeds: A review. *Food Res. Int.* **2017**, *101*, 1–16. [CrossRef]

41. Awika, J.M.; Rooney, L.W. Sorghum phytochemicals and their potential impact on human health. *Phytochemistry* **2004**, *65*, 1199–1221. [CrossRef]

42. Ghosh, D. Tannins from foods to combat diseases. *Int. J. Pharm. Res. Rev.* **2015**, *4*, 40–44.

43. Adebo, O.A. Metabolomics, Physicochemical Properties and Mycotoxin Reduction of Whole Grain *ting* (a Southern African Fermented Food) Produced via Natural and Lactic Acid Bacteria (LAB) Fermentation. Ph.D. Thesis, University of Johannesburg, Johannesburg, South Africa, October 2018.

44. Goel, G.; Puniya, A.K.; Aguliar, C.N.; Singh, K. Interaction of gut microflora with tannins in feeds. *Naturwissenschaften* **2005**, *92*, 497–503. [CrossRef]

45. Clifford, M.N. Chlorogenic acids and other cinnamates-nature, occurrence, and dietary burden. *J. Sci. Food Agric.* **1999**, *79*, 362–372. [CrossRef]

46. Bondia-Pons, I.; Aura, A.M.; Vuorela, S.; Kolehmainen, M.; Mykkanen, H.; Poutanen, K. Rye phenolics in nutrition and health. *J. Cereal Sci.* **2009**, *49*, 323–336. [CrossRef]

47. Shadidi, F.; Chandrasekara, A. Millet grain phenolics and their role in disease risk reduction and health promotion: A review. *J. Funct. Food* **2013**, *5*, 570–581. [CrossRef]

48. Shadidi, F.; Yeo, J. Insoluble-bound phenolics in food. *Molecules* **2016**, *21*, 1216. [CrossRef]

49. Gong, L.; Cao, W.; Chi, H.; Wang, J.; Zhang, H.; Liu, J.; Sun, B. Whole cereal grains and potential health effects: Involvement of the gut microbiota. *Food Res. Int.* **2018**, *103*, 84–102. [CrossRef]

50. Kumar, N.; Goel, N. Phenolic acids: Natural versatile molecules with promising therapeutic applications. *Biotechnol. Rep.* **2019**, *24*, e00370. [CrossRef]

51. Masebe, K.M.; Adebo, O.A. Production and quality characteristics of a probiotic beverage from watermelon (Citrullus lanatus). In *Engineering, Technology and Waste Management (SETWM-19), Proceedings of the 17th Johannesburg International Conference on Science, Johannesburg, South Africa, 18–19 November 2019;* Fosso-Kankeu, E., Waanders, F., Bulsara, H.K.P., Eds.; Eminent Association of Pioneers and North-West University: Johannesburg, South Africa, 2019; pp. 42–49.

52. Marsh, A.J.; Hill, C.; Ross, R.P.; Cotter, P.D. Fermented beverages with health-promoting potential: Past and future perspectives. *Trend Food Sci. Technol.* **2014**, *38*, 113–124. [CrossRef]

53. Adebo, O.A.; Njobeh, P.B.; Adebiyi, J.A.; Gbashi, S.; Phoku, J.Z.; Kayitesi, E. Fermented pulse-based foods in developing nations as sources of functional foods. In *Functional Food-Improve Health through Adequate Food;* Hueda, M.C., Ed.; InTech: Rijeka, Croatia, 2017; pp. 77–109.

54. Kaushik, G.; Satya, S.; Naik, S.N. Food processing a tool to pesticide residue dissipation—A review. *Food Rev. Int.* **2009**, *42*, 26–40. [CrossRef]

55. Adebiyi, J.A.; Njobeh, P.B.; Kayitesi, E. Assessment of nutritional and phytochemical quality of *Dawadawa* (an African fermented condiment) produced from Bambara groundnut (*Vigna subterranea*). *Microchem. J.* **2019**, *149*, 104034. [CrossRef]

56. Llopart, E.E.; Drago, S.R.; De Greef, D.M.; Torres, R.L.; Gonzalez, A.R. Effects of extrusion conditions on physical and nutritional properties of extruded whole grain red sorghum (*Sorghum* spp.). *Int. J. Food Sci. Nutr.* **2014**, *65*, 34–41. [CrossRef]

57. Kantor, L.S.; Variyam, J.N.; Allshouse, J.E.; Putnam, J.J.; Lin, B.H. Choose a variety of grains daily, especially whole grains: A challenge for consumers. *J. Nutr.* **2001**, *131*, 473S–486S. [CrossRef]

58. Kyrø, C.; Skeie, G.; Dragsted, L.O.; Christensen, J.; Overvad, K.; Hallmans, G.; Johansson, I.; Lund, E.; Slimani, N.; Johnsen, J.F.; et al. Intake of whole grain in Scandinavia: Intake, sources and compliance with new national recommendations. *Scand J. Public Health* **2012**, *40*, 76–84. [CrossRef]

59. Eklund-Jonsson, C.; Sanderg, A.S.; Alminger, M.L. Reduction of phytate content while preserving minerals during whole grain cereal *tempe* fermentation. *J. Cereal Sci.* **2006**, *44*, 154–160. [CrossRef]

60. Oyarekua, M.A. Effect of co-fermentation on nutritive quality and pasting properties of maize/cowpea/sweet potato as complementary food. *Afr. J. Food Agric. Nutr. Dev.* **2013**, *13*, 7171–7191. [CrossRef]

61. Obinna-Echem, P.C.; Beal, P.; Kuri, V. Effect of processing method on the mineral content of Nigerian fermented maize infant complementary food-*akamu*. *LWT-Food Sci. Technol.* **2015**, *61*, 145–151. [CrossRef]

62. Salar, R.J.; Purewal, S.S.; Bhatti, M.S. Optimization of extraction conditions and enhancement of phenolic content and antioxidant activity of pearl millet fermented with *Aspergillus awamori* MTCC-54. *Resour. Effic. Technol.* **2016**, *2*, 148–157. [CrossRef]

63. Di Stefano, E.; White, J.; Seney, S.; Hekmat, S.; McDowell, T.; Sumarah, M.; Reid, G. A novel millet-based probiotic fermented food for the developing world. *Nutrients* **2017**, *9*, 529. [CrossRef]

64. Sheela, P.; UmaMaheswari, T.; Kanchana, S.; Kamalasundari, S.; Hemalatha, G. Development and evaluation of fermented millet milk based curd. *J. Pharmacog. Phytochem.* **2018**, *7*, 714–717.

65. Wang, Y.; Compaoré-Sérémé, D.; Sawadogo-Lingani, H.; Coda, R.; Katina, K.; Maina, N.J. Influence of dextran synthesized in situ on the rheological, technological and nutritional properties of whole grain pearl millet bread. *Food Chem.* **2019**, *285*, 221–230. [CrossRef]

66. Wu, H.; Rui, X.; Li, W.; Xiao, Y.; Zhou, J.; Dong, M. Whole-grain oats (*Avena sativa* L.) as a carrier of lactic acid bacteria and a supplement rich in angiotensin I-converting enzyme inhibitory peptides through solid-state fermentation. *Food Funct.* **2018**, *9*, 2270–2281. [CrossRef]

67. Gamel, T.H.; Abdel-Aal, E.M.; Tosh, S.M. Effect of yeast-fermented and sour-dough making processes on physicochemical characteristics of β-glucan in whole wheat/oat bread. *LWT-Food Sci. Technol.* **2015**, *60*, 78–85. [CrossRef]

68. Zannini, E.; Jeske, S.; Lynch, K.M.; Arendt, E.K. Development of novel quinoa-based yoghurt fermented with dextran producer *Weissella cibaria* MG1. *Int. J. Food Microbiol.* **2018**, *268*, 19–26. [CrossRef]

69. Ayyash, M.; Johnson, S.K.; Liu, S.Q.; Al-Mheiri, A.; Abushelaibi, A. Cytotoxicity, antihypertensive, antidiabetic and antioxidant activities of solid-state fermented lupin, quinoa and wheat by Bifidobacterium species: In vitro investigations. *LWT-Food Sci. Technol.* **2018**, *95*, 295–302. [CrossRef]

70. Buddrick, O.; Jones, O.A.H.; Cornell, H.J.; Small, D.M. The influence of fermentation processes and cereal grains in wholegrain bread on reducing phytate content. *J. Cereal Sci.* **2014**, *59*, 3–8. [CrossRef]

71. Koistinen, V.M.; Mattila, O.; Katina, K.; Poutanen, K.; Aura, A.M.; Hanhineva, K. Metabolic profiling of sourdough fermented wheat and rye bread. *Sci. Rep.* **2018**, *8*, 5684. [CrossRef]

72. Johansson, D.P.; Lee, I.; Risérus, U.; Langton, M.; Landberg, R. Effects of unfermented and fermented whole grain rye crisp breads served as part of a standardized breakfast, on appetite and postprandial glucose and insulin responses: A randomized cross-over trial. *PLoS ONE* **2015**, *10*, e0122241. [CrossRef]

73. Raninen, K.; Lappi, J.; Kolehmainen, M.; Kolehmainen, M.; Mykkanen, H.; Poutanen, K.; Raatikainen, O. Diet-derived changes by sourdough-fermented rye bread in exhaled breath aspiration ion mobility spectrometry profiles in individuals with mild gastrointestinal symptoms. *Int. J. Food Sci. Nutr.* **2017**, *8*, 987–996. [CrossRef]

74. Lee, I.; Shi, L.; Webb, D.L.; Hellström, P.M.; Risérus, U.; Landberg, R. Effects of whole-grain rye porridge with added inulin and wheat gluten on appetite, gut fermentation and postprandial glucose metabolism: A randomised, cross-over, breakfast study. *Br. J. Nutr.* **2017**, *116*, 2139–2149. [CrossRef]

75. Beckmann, M.; Lloyd, A.J.; Haldar, S.; Seal, C.; Brandt, K.; Draper, J. Hydroxylated phenylacetamides derived from bioactive benzoxazinoids are bioavailable in humans after habitual consumption of whole grain sourdough rye bread. *Mol. Nutr. Food Res.* **2013**, *57*, 1859–1873. [CrossRef]

76. Zamaratskaia, G.; Johansson, D.P.; Junqueira, M.A.; Deissler, L.; Langton, M.; Hellstrom, P.M.; Landberg, R. Impact of sourdough fermentation on appetite and postprandial metabolic responses—A randomised cross-over trial with whole grain rye crispbread. *Br. J. Nutr.* **2017**, *118*, 686–697. [CrossRef]

77. Ikediobi, C.O.; Olugboji, O.; Okoh, P.N. Cyanide profile of component parts of sorghum (*Sorghum bicolor* L. Moench) sprouts. *Food Chem.* **1988**, *27*, 167–175. [CrossRef]

78. Ragaee, S.; Abdel-Aal, E.M. Pasting properties of starch and protein in selected cereals and quality of their food products. *Food Chem.* **2006**, *95*, 9–18. [CrossRef]

79. Dlamini, N.R.; Taylor, J.R.N.; Rooney, L.W. The effect of sorghum type and processing on the antioxidant properties of African sorghum-based foods. *Food Chem.* **2007**, *105*, 1412–1419. [CrossRef]

80. Taylor, J.; Taylor, J.R.N. Protein biofortified sorghum: Effect of processing into traditional African foods on their protein quality. *J. Agric. Food Chem.* **2011**, *59*, 2386–2392. [CrossRef]

81. Akingbala, J.O.; Rooney, L.W.; Fauboin, J.M. Physical, chemical, and sensory evaluation of *ogi* from sorghum of differing kernel characteristics. *J. Food Sci.* **1981**, *76*, 1532–1536. [CrossRef]

82. Mukuru, S.Z.; Butler, L.G.; Rogler, J.C.; Kirleis, A.W.; Ejeta, G.; Axtell, J.D.; Mertz, E.T. Traditional processing of high-tannin sorghum grain in Uganda and its effect on tannin, protein digestibility, and rat growth. *J. Agric. Food Chem.* **1992**, *40*, 1172–1175. [CrossRef]

83. Kruger, J.; Taylor, J.R.N.; Oelofse, A. Effects of reducing phytate content in sorghum through genetic modification and fermentation on in vitro iron availability in whole grain porridges. *Food Chem.* **2012**, *131*, 220–224. [CrossRef]

84. Adebo, O.A.; Njobeh, P.B.; Adebiyi, J.A.; Kayitesi, E. Co-influence of fermentation time and temperature on physicochemical properties, bioactive components and microstructure of *ting* (a Southern African food) from whole grain sorghum. *Food Biosci.* **2018**, *25*, 118–127. [CrossRef]

85. Adebo, O.A.; Njobeh, P.B.; Kayitesi, E. Fermentation by *Lactobacillus fermentum* strains (singly and in combination) enhances the properties of *ting* from two whole grain sorghum types. *J. Cereal Sci.* **2018**, *82*, 49–56. [CrossRef]

86. Tamene, A.; Kariluoto, S.; Baye, K.; Humblot, C. Quantification of folate in the main steps of traditional processing of tef injera, a cereal based fermented staple food. *J. Cereal Sci.* **2019**, *87*, 225–230. [CrossRef]

87. Gotcheva, V.; Pandiella, S.S.; Angelov, A.; Roshkova, Z.; Webb, C. Monitoring the fermentation of the traditional Bulgarian beverage *boza*. *Int. J. Food Sci. Technol.* **2001**, *36*, 129–134. [CrossRef]

88. Mustafa, K.D.; Adem, E. Comparison of autoclave, microwave, IR and UV-stabilization of whole wheat flour branny fractions upon the nutritional properties of whole wheat bread. *J. Food Sci. Technol.* **2014**, *51*, 59–66.

89. Struyf, N.; Laurent, J.; Verspreet, J.; Verstrepen, K.J.; Coutrin, C.M. *Saccharomyces cerevisiae* and *Kluyveromyces marxianus* cocultures allow reduction of fermentable oligo-, di-, and monosaccharides and polyols levels in whole wheat bread. *J. Agric. Food Chem.* **2017**, *65*, 8704–8713. [CrossRef] [PubMed]

90. García-Mantrana, I.; Yebra, M.J.; Haros, M.; Monedero, V. Expression of bifidobacterial phytases in *Lactobacillus casei* and their application in a food model of whole-grain sourdough bread. *Int. J. Food Microbiol.* **2016**, *216*, 18–24. [CrossRef] [PubMed]

91. Dey, T.B.; Kuhad, R.C. Enhanced production and extraction of phenolic compounds from wheat by solid-state fermentation with *Rhizopus oryzae* RCK2012. *Biotechnol. Rep.* **2014**, *4*, 120–127.

92. Starzyńska-Janiszewska, A.; Stodolak, B.; Socha, R.; Mickowska, B.; Wywrocka-Gurgul, A. Spelt wheat *tempe* as a value-added whole-grain food product. *LWT-Food Sci. Technol.* **2019**, *113*, 108250. [CrossRef]

93. Nuobariene, L.; Cizeikiene, D.; Gradzeviciute, E.; Hansen, A.S.; Rasmussen, S.K.; Juodeikiene, G.; Vogensen, F.K. Phytase-active lactic acid bacteria from sourdoughs: Isolation and identification. *LWT-Food Sci. Technol.* **2015**, *63*, 766–772. [CrossRef]

94. Adebo, O.A.; Njobeh, P.B.; Mulaba-Bafubiandi, A.F.; Adebiyi, J.A.; Desobgo, Z.S.C.; Kayitesi, E. Optimization of fermentation conditions for *ting* production using response surface methodology. *J. Food Proc. Preserv.* **2018**, *42*, e13381. [CrossRef]

95. Salar, R.K.; Certik, M.; Brezova, V. Modulation of phenolic content and antioxidant activity of maize by solid state fermentation with *Thamnidium elegans* CCF 1456. *Biotechnol. Bioprocess Eng.* **2012**, *17*, 109–116. [CrossRef]

96. Hole, A.S.; Rud, I.; Grimmer, S.; Sigl, S.; Narvhus, J.; Sahlstrøm, S. Improved bioavailability of dietary phenolic acids in whole grain barley and oat groat following fermentation with probiotic *Lactobacillus acidophilus*, *Lactobacillus johnsonii*, and *Lactobacillus reuteri*. *J. Agric. Food Chem.* **2012**, *60*, 6369–6375. [CrossRef]

97. Liukkonen, K.H.; Katina, K.; Wilhelmsson, A.; Myllymaki, O.; Lampi, A.M.; Kariluoto, S.; Piironen, V.; Heinonen, S.M.; Nurmi, T.; Adlercreutz, H.; et al. Process-induced changes on bioactive compounds in whole grain rye. *Proc. Nutr. Soc.* **2003**, *62*, 117–122. [CrossRef]

98. Katina, K.; Liukkonen, K.H.; Kaukovirta-Norja, A.; Adlercreutz, H.; Heinonen, S.M.; Lampi, A.M.; Pihlava, J.M.; Poutanen, K. Fermentation-induced changes in the nutritional value of native or germinated rye. *J. Cereal Sci.* **2007**, *46*, 348–355. [CrossRef]

99. Skrajda-Brdak, M.; Konopka, I.; Tańska, M.; Czaplicki, S. Changes in the content of free phenolic acids and antioxidative capacity of wholemeal bread in relation to cereal species and fermentation type. *Eur. Food Res. Technol.* **2019**, *245*, 2247–2256. [CrossRef]

100. Mohapatra, D.; Patel, A.S.; Kar, A.; Deshpande, S.S.; Tripathi, M.K. Effect of different processing conditions on proximate composition, anti-oxidants, anti-nutrients and amino acid profile of grain sorghum. *Food Chem.* **2019**, *271*, 129–135. [CrossRef] [PubMed]

101. Đordevic, T.M.; Šiler-Marinkovic, S.S.; Dimitrijevic-Brankovic, S.I. Effect of fermentation on antioxidant properties of some cereals and pseudo cereals. *Food Chem.* **2010**, *119*, 957–963. [CrossRef]

102. Ripari, V.; Bai, Y.; Gänzle, M.G. Metabolism of phenolic acids in whole wheat and rye malt sourdoughs. *Food Microbiol.* **2019**, *77*, 43–51. [CrossRef]

103. Dey, T.B.; Kuhad, R.C. Upgrading the antioxidant potential of cereals by their fungal fermentation under solid-state cultivation conditions. *Lett. Appl. Microbiol.* **2014**, *59*, 493–499.

104. Gänzle, M.G. Enzymatic and bacterial conversions during sourdough fermentation. *Food Microbiol.* **2014**, *37*, 2–10. [CrossRef]

105. Katina, K.; Juvonen, R.; Laitila, A.; Flander, L.; Nordlund, E.; Kariluoto, S.; Piironen, V.; Poutanen, K. Fermented wheat bran as a functional ingredient in baking. *Cereal Chem.* **2012**, *89*, 126–134. [CrossRef]

106. Czerny, M.; Schieberie, P. Important aroma compounds in freshly ground whole meal and white wheat flour-identification and quantitative changes during sourdough fermentation. *J. Agric. Food Chem.* **2002**, *50*, 6835–6840. [CrossRef]

107. Rodríguez, H.; Curiel, J.A.; Landete, J.M.; de las Rivas, B.; de Felipe, F.L.; Gomez-Cordoves, C.; Mancheno, J.M.; Munoz, R. Food phenolics and lactic acid bacteria. *Int. J. Food Microbiol.* **2009**, *132*, 79–90. [CrossRef]

108. Svensson, L.; Sekwati-Monang, B.; Lutz, D.L.; Schieber, A.; Gänzle, M.G. Phenolic acids and flavonoids in nonfermented and fermented red sorghum (*Sorghum bicolor* (L.) Moench). *J. Agric. Food Chem.* **2010**, *58*, 9214–9220. [CrossRef] [PubMed]

109. Sánchez-Maldonado, A.F.; Schieber, A.; Gänzle, M.G. Structure-function relationships of the antibacterial activity of phenolic acids and their metabolism by lactic acid bacteria. *J. Appl. Microbiol.* **2011**, *111*, 1176–1184. [CrossRef] [PubMed]

110. Raccach, M. The antimicrobial activity of phenolic antioxidant in food: A review. *J. Food Saf.* **1984**, *6*, 141–170. [CrossRef]

111. Filannino, P.; Bai, Y.; Di Cagno, R.; Gobbetti, M.; Gänzle, M.G. Metabolism of phenolic compounds by *Lactobacillus* spp. during fermentation of cherry juice and broccoli puree. *Food Microbiol.* **2015**, *46*, 272–279. [CrossRef] [PubMed]

112. Gaur, G.; Oh, J.H.; Filannino, P.; Gobbetti, M.; van Pijkeren, J.P.; Gänzle, M.G. Genetic determinants of hydroxycinnamic acid metabolism in heterofermentative lactobacilli. *Appl. Environ. Microbiol.* **2019**. [CrossRef] [PubMed]

113. Kurutas, E.B. The importance of antioxidants which play the role in cellular response against oxidative/nitrosative stress: Current state. *Nutr. J.* **2016**, *15*, 71. [CrossRef] [PubMed]

114. Slavin, J. Why whole grains are protective: Biological mechanisms. *Proc. Nutr. Soc.* **2003**, *62*, 129–134. [CrossRef]

115. Verma, B.; Hucl, P.; Chibbar, R.N. Phenolic content and antioxidant properties of bran in 51 wheat cultivars. *Cereal Chem.* **2008**, *85*, 544–549. [CrossRef]

116. Sevgi, K.; Tepe, B.; Sarikurkcu, C. Antioxidant and DNA damage protection potentials of selected phenolic acids. *Food Chem. Toxicol.* **2015**, *77*, 12–21. [CrossRef]

117. Wattenberg, L.W. Chemoprevention of cancer. *Cancer Res.* **1985**, *45*, 1–8. [CrossRef]

118. Slavin, J.L. Mechanisms for the impact of whole grain foods on cancer risk. *J. Am. Coll. Nutr.* **2000**, *19*, 300S–307S. [CrossRef] [PubMed]

119. Ghasemzadeh, A.; Ghasemzadeh, N. Flavonoids and phenolic acids: Role and biochemical activity in plants and human. *J. Med. Plants Res.* **2011**, *5*, 6697–6703. [CrossRef]

120. Juurlink, B.H.J.; Azouz, H.J.; Aldalati, A.M.Z.; AlTinawi, B.M.H.; Ganguly, P. Hydroxybenzoic acid isomers and the cardiovascular system. *Nutr. J.* **2014**, *13*, 63. [CrossRef] [PubMed]

121. Beta, T.; Nam, S.; Dexter, J.E.; Sapirstein, H.D. Phenolic content and antioxidant activity of pearled wheat and roller-mill fractions. *Cereal Chem.* **2005**, *82*, 390–393. [CrossRef]

122. Kim, K.H.; Tsao, R.; Yang, R.; Cui, S.W. Phenolic acid profiles and antioxidant activities of wheat bran extracts and the effect of hydrolysis conditions. *Food Chem.* **2006**, *95*, 466–473. [CrossRef]

123. Li, L.; Shewry, P.R.; Ward, J.L. Phenolic acids in wheat varieties in the HEALTHGRAIN diversity screen. *J. Agric. Food Chem.* **2008**, *56*, 9732–9739. [CrossRef]

124. Khan, R.A.; Khan, M.R.; Sahreen, S.; Ahmed, M. Evaluation of phenolic contents and antioxidant activity of various solvent extracts of *Sonchus asper* (L.) Hill. *Chem. Central J.* **2012**, *6*, 12. [CrossRef]

125. Sun, T.; Ho, C.T. Antioxidant activities of buckwheat extracts. *Food Chem.* **2005**, *90*, 743–749. [CrossRef]

126. Dai, J.; Mumper, R. Plant phenolics: Extraction, analysis and their antioxidant and anticancer properties. *Molecules* **2010**, *15*, 7313–7352. [CrossRef]

127. Žilić, S.; Serpen, A.; Akıllıoğlu, G.; Janković, M.; Gökmen, V. Distributions of phenolic compounds, yellow pigments and oxidative enzymes in wheat grains and their relation to antioxidant capacity of bran and debranned flour. *J. Cereal Sci.* **2012**, *56*, 652–658. [CrossRef]

128. Benedetti, S.; Primiterra, M.; Tagliamonte, M.C.; Carnevali, A.; Gianotti, A.; Bordoni, A.; Canestrari, F. Counteraction of oxidative damage in the rat liver by an ancient grain (Kamut brand khorasan wheat). *Nutrition* **2012**, *28*, 436–441. [CrossRef] [PubMed]

129. Alam, M.N.; Bristi, N.J.; Rafiquzzaman, M. Review on in vivo and in vitro methods evaluation of antioxidant activity. *Saudi Pharm. J.* **2013**, *21*, 143–152. [CrossRef] [PubMed]

130. Verni, M.; Verardo, V.; Rizzello, C.G. How fermentation affects the antioxidant properties of cereals and legumes. *Foods* **2019**, *8*, 362. [CrossRef] [PubMed]

131. Gianotti, A.; Danesi, F.; Verardo, V.; Serrazanetti, D.I.; Valli, V.; Russo, A.; Riciputi, Y.; Tossani, N.; Caboni, M.F.; Guerzoni, M.E.; et al. Role of whole grain in the in vivo protection from oxidative stress. *Front. Biosci.* **2011**, *16*, 1609–1618. [CrossRef] [PubMed]

132. Moore, J.; Cheng, Z.; Hao, J.; Guo, G.; Liu, J.G.; Lin, C.; Yu, L.L. Effects of solid state yeast treatment on the antioxidant properties and protein and fiber compositions of common hard wheat bran. *J. Agric. Food Chem.* **2007**, *55*, 10173–10182. [CrossRef] [PubMed]

133. Combest, S.; Warren, C. Perceptions of college students in consuming whole grain foods made with brewers' spent grain. *Food Sci. Nutr.* **2018**, *7*, 225–237. [CrossRef]

134. Kuznesof, S.; Brownlee, I.A.; Moore, C.; Richardson, D.P.; Jebb, S.A.; Seal, C.J. WHOLEheart study participant acceptance of wholegrain foods. *Appetite* **2012**, *59*, 187–193. [CrossRef]

135. Kamar, M.; Evans, C.; Hugh-Jones, S. Factors influencing adolescent whole grain intake: A theory-based qualitative study. *Appetite* **2016**, *101*, 125–133. [CrossRef]

136. Chea, M.; Mobley, A.R. Factors associated with identification and consumption of whole-grain foods in a low-income population. *Curr. Dev. Nutr.* **2019**, *3*, nzz064. [CrossRef]

Article

Microwave Hydrodiffusion and Gravity (MHG) Extraction from Different Raw Materials with Cosmetic Applications

Lucía López-Hortas [1], Elena Falqué [2] , Herminia Domínguez [1] and María Dolores Torres [1,*]

1   Department of Chemical Engineering, Faculty of Sciences, University of Vigo, Edificio Politécnico, As Lagoas s/n, 32004 Ourense, Spain; luloho@gmail.com (L.L.-H.); herminia@uvigo.es (H.D.)
2   Department of Analytical Chemistry, Faculty of Sciences, University of Vigo, Edificio Politécnico, As Lagoas s/n, 32004 Ourense, Spain; efalque@uvigo.es
*   Correspondence: matorres@uvigo.es

Received: 4 December 2019; Accepted: 22 December 2019; Published: 25 December 2019

Abstract: Microwave hydrodiffusion and gravity (MHG) and ethanolic solid-liquid extraction were compared using selected plant sources. Their bioactive profile, color features, and proximate chemical characterization were determined. MHG extracts, commercial antioxidants, and three distinct types of thermal spring water were used in a sunscreen cream formulation. Their bioactive capacity, chemical and rheological properties were evaluated. MHG *Cytisus scoparius* flower extract provided the highest bioactive properties. *Pleurotus ostreatus* MHG liquor exhibited the highest total solid extraction yield. The *Brassica rapa* MHG sample stood out for its total protein content and its monosaccharide and oligosaccharide concentration. *Quercus robur* acorns divided into quarters supplied MHG extract with the lowest energy requirements, highest DPPH inhibition percentage, total lipid content and the highest enzyme inhibition. The chemical and bioactive capacities stability of the sunscreen creams elaborated with the selected MHG extracts and the thermal spring waters showed a similar behavior than the samples containing commercial antioxidants.

Keywords: *Cytisus scoparius*; *Pleurotus ostreatus*; *Brassica rapa*; *Quercus robur*; sun creams; thermal spring waters

---

1. Introduction

In the last decade, cosmetics from natural botanical sources have gained increasing interest due to their healthy features avoiding synthetic components in the matrices [1,2]. The consumer demand has notably grown due to the rise of sensitive skin, atopic diseases or allergies. In this field, spring waters can be used in different treatments and rehabilitation of patients in the hydrotherapy context. The positive effect of the hot spring water therapy in functional improvements, musculoskeletal diseases, and rehabilitation of patients in sports medicine is well known [3]. Recent studies have indicated that the physical properties and chemical effects of this type of water have a potential application in immuno-inflammatory processes, chronic pain diseases, chronic cardiac procedures, metabolic syndromes and neurological illnesses [4]. The use of thermal waters in dermatology treatments provides therapeutic benefits due to their non-pathogenic microbes and mineral content. The use of different topical products made with thermal waters enhances several dermatitis diseases [5–7].

The natural bioactive compounds used in cosmetics can provide a variety of activities, acting as humectants, photoprotective, antioxidant, antimicrobial, or anti-aging agents [8–10]. Alternative botanical sources of interest could be employed in cosmetic products since numerous bioactive compounds with health benefits in this sense derived from vegetable raw materials. In this work, some undervalued wild resources and agriculture crops were used to recover extracts with probable use in cosmetic products. Extracts of *Cytisus scoparius* flowers were employed in a sunscreen cream since this plant contains polyphenolic compounds with antioxidant and antibacterial capacities without skin irritation response [11,12]. *Pleurotus ostreatus* is a mushroom widely used in cosmeceuticals products. This cosmetic ingredient offers antioxidant, antimicrobial and anti-tyrosinase activities [13,14]. *Brassica rapa* tap root extracts also have a potential cosmetic use since its tyrosinase enzymatic inhibition promotes the skin depigmentation [15,16]. For this reason, the authors consider the evaluation of the *Brassica rapa* var. *rapa* leaves extract since this crop is a traditional and extensive plant produced in Galicia area (northwest of Spain). *Quercus robur* is a tree very abundant in the Atlantic forest. Their acorns presented a noticeable content of antioxidant and phenolic compounds that can be very useful in cosmetic products [17,18]. It should be highlighted that despite the number of natural cosmetic applications, further scientific studies with comprehensive physicochemical and mechanical analysis of the natural sunscreen creams are required in order to improve the processing, storage and quality of the end product.

The use of environmentally friendly methods for the extraction of the functional ingredients of cosmetics has been another challenge in recent years [19]. Microwave hydrodiffusion and gravity (MHG) could be a solvent-free extraction attractive alternative to obtain bioactive fractions from natural resources in a sustainable manner [20] in contrast with the conventional solid-liquid extraction (SLE) technique. For example, *C. scoparius* branches were extracted with *n*-hexane solvent in a Soxhlet extractor [11]. These types of extraction procedures demand important volumes of solvents with high purity and extensive time extraction period which means that these processes are expensive and imply problems associated with the toxic danger features of the used solvents for foodstuff, pharmacology and cosmetic uses [21].

In this context, the aim of this paper was the employment of MHG technology with different agroforestry raw materials for their valorization as a source of high value-added antioxidant compounds. Yields and energy consumptions were evaluated. Total phenolic content, antioxidant capacity, total carotenoid content and capacity of anti-elastase and -tyrosinase enzymatic inhibition of the collected liquid phases by this innovative extraction method were evaluated. Principal macronutrients present in these samples as well as their color features were also examined. These properties were also defined for the recovered solid phases. This study was complemented with the incorporation of the selected extracts in a sun cream product formulated with mineral waters from different thermal spring waters. These cosmetics were assayed to determine their physicochemical parameters and rheological characteristics. An accelerated oxidation test was realized to determine their oxidation stability.

2. Materials and Methods

2.1. Collection of Samples and Reagents

Cytisus scoparius flowers and *Quercus robur* L. acorns without cupules were manually collected in Outeiro de Rei (Lugo, Spain), whereas *Pleurotus ostreatus* mushrooms and *Brassica rapa* var. *rapa* leaves were supplied for a local grocery store (Ourense, Spain). All samples were cold stored ($-18 \pm 2\,°C$) until further use, apart from *Q. robur* which was maintained at room temperature. Prior to the MHG process, frozen samples were defrosted at $4 \pm 2\,°C$ for 24 h whereas acorns were manually chopped with a longitudinal incision and divided into quarters. In all cases, the employed reagents were of analytical grade.

2.2. Extraction Procedures

Microwave hydrodiffusion and gravity was used to extract the bioactive fractions of tested raw materials. For this purpose, the extractions were made on a multimode microwave extractor (NEOS-GR, Milestone Srl, Italy) as reported elsewhere [22]. Samples (100 g) were placed on an open vessel (1.5 L) and submitted at a power density of 1 W/g. These conditions were selected based on the results previously reported for other similar materials [23]. The collected volume (each 5 mL), time and temperature were recorded in all cases. Note that the MHG process was ended whenever it was not possible to collect more liquid extract. Liquors were drained by gravity on a condenser outside the microwave. Extracts were cold-stored at 4 ± 2 °C in the absence of light until further analysis. Experiments were made at least in duplicate.

An approximation of the extraction procedure's environmental influences [24] and energy requirements [25] were calculated as previously reported. In the first case, the required dioxide carbon released was determined assuming 800 g of CO_2 by the consumed of 1 kWh from coal or fuel. Concerning energy consumption, it was calculated as the necessary time by the device power necessary for each trial.

For comparative purposes, samples (1 g) were submitted to solid-liquid extraction (SLE) with ethanol solution (10 mL) (Sigma-Aldrich Corp., St. Louis, MO, USA). Note here that *B. rapa* leaves were treated with 90% (*v/v*) ethanol solution whereas the remaining samples were extracted with 70% (*v/v*) ethanol solution in accordance with previous assays. This solvent extraction was performed at 150 rpm and 40 ± 2 °C for 24 h in an orbital incubator shaker (Innova 4000, New Brunswick Scientific, Edison, NJ, USA) in the absence of light conditions. Collected extracts were filtered and kept at 4 ± 2 °C in darkness until further analysis.

2.3. Extracts Characterization

The procedures used to characterize the collected liquid extracts extracted by MHG and by SLE procedures are presented in the following sections.

2.3.1. pH

pH values were determined using a pH meter (GLP 21, Hach Lange Spain, S.L.U., Barcelona, Spain) at room temperature. Previously, a calibration was realized with the corresponding standard solutions.

2.3.2. Total Solid Content

Total solid extraction yield (g extract/g raw material dry weight) was calculated based on the initial moisture of the raw materials utilized, the collected volume of the different extracts and the total solid content (g dry residue/L) determined gravimetrically with an extract aliquot (1 mL) by oven drying at 105 °C during 24 h.

2.3.3. Bioactive Profile

The total phenolic content of the tested liquid extracts was made by the Folin–Ciocalteu test [26], expressing the obtained content as milligrams of gallic acid equivalents (GAE). For this test, samples (0.50 mL) or the standard (gallic acid) (Sigma-Aldrich Corp., St. Louis, MO, USA) was incorporated to distilled water (3.75 mL), Folin–Ciocalteu reagent (0.25 mL, 1:1, *v/v*) (Panreac Química, S.L.U., Castellar del Vallès, Barcelona, Spain) and sodium carbonate solution (0.50 mL, 10%, *w/v*) (Sigma-Aldrich Corp., St. Louis, Missouri, USA). After keeping the mixtures in the absence of light at room temperature for about 60 min, the absorbance measurements were made at 765 nm.

Concerning the antioxidant profile, the ABTS (2,2′-azino-bis(3-ethylbenzothiazoline-6-sulfonic acid) diammonium salt radical cation scavenging capacity was also determined [27], expressing the results as Trolox equivalents antioxidant capacity (TEAC). In this case, samples (20 µL) or the standard (Trolox) (Sigma-Aldrich Corp., St. Louis, MO, USA) were mixed with diluted $ABTS^+$ solution (2.00 mL) (Sigma-Aldrich Corp., St. Louis, MO, USA). Mixtures were maintained at 30 ± 2 °C for 6 min before reading the absorbance at 734 nm.

The reduction of the ferric 2,4,6-tripyridyl-s-triazine (TPTZ) complex was conducted for the above samples using the ferric reducing antioxidant power (FRAP) method [28]. Note that the ascorbic acid (Merck KGaA, Darmstadt, Germany) and iron (II) sulphate heptahydrate (Sigma-Aldrich Corp.,

St. Louis, MO, USA) were employed as standards. Liquid extracts of 100 µL were added to 3.00 mL of the FRAP reagent. After 6 min at room temperature, the absorbance was read at 593 nm.

The DPPH (α,α-diphenyl-β-picrylhydrazyl) radical scavenging activity of the above liquors was also determined [29]. Samples (50 µL) were added to the DPPH radical solution (2 mL) (Sigma-Aldrich Corp., St. Louis, MO, USA). Afterward, the absorbance decrease was recorded at 515 nm from 0 to 16 min using a blank sample (distilled water).

The determination of the total carotenoid content of the different extracts was realized following Khosa et al. [30] method with some modifications. In few words, 1 ± 0.01 g of each sample was submitted at a liquid-liquid extraction with 50 mL of *n*-hexane (Panreac Química, S.L.U., Castellar del Vallès, Barcelona, Spain)/acetone (Carlo Erba Reagents, S.A., Sabadell, Barcelona, Spain)/absolute ethanol (Sigma-Aldrich Corp., St. Louis, Missouri, USA) (2:1:1, *v/v/v*) for 20 min at 200 rpm in darkness conditions using an orbital shaker (OL30-ME, Ovan, Barcelona, España), then was centrifuged at 4000 rpm for 10 min and measured at 420 nm. β-carotene (Thermo Fisher Scientific, Geel, Belgium) was utilized as standard and results were expressed as µg β-carotene/g raw material dry weight.

All above analytical determinations were carried out on a spectrophotometer Hitachi U-2000 (Tokyo, Japan) at least in triplicate. This point was also applicable to the next spectrophotometric methods.

2.3.4. Capacity of Elastase and Tyrosinase Enzymatic Inhibition

The capacity of elastase and tyrosinase enzymatic inhibition of the studied extracted samples were conducted by an external service following the indications reported with slight modifications by Liyanaarachchi et al. [31] and Chiari et al. [32], respectively.

2.3.5. Macronutrients Measurements

MHG and SLE extracts were subjected to certain analyses to determine their main components. Their monosaccharide composition was assayed by high-performance liquid chromatography (HPLC) following a modified Balboa et al. [33] method using an Agilent equipment with a differential refractive index detector. Aminex HPX-87H and Aminex HPX-87P columns (300 × 7.8 mm, Bio-Rad Laboratories, S.A., Madrid, Spain) were employed by the separation of the different carbohydrate compounds eluted with 3 mM H_2SO_4 (Merck KGaA, Darmstadt, Germany) at 0.6 mL/min operating at 50 ± 2 °C and deionized water at 0.4 mL/min operating at 80 ± 2 °C, respectively. Samples were neutralized with barium carbonate (Thermo Fisher Scientific, Geel, Belgium) before analysis on the Aminex HPX-87P column. Their oligosaccharide composition was determined following the Garrote et al. [34] method and consequently, samples (2.4 ± 0.10 g) were submitted to a previous hydrolysis treatment with 4% (*w/w*) sulphuric acid (Merck KGaA, Darmstadt, Germany) at 121 ± 2 °C during 40 min. The estimated concentrations were calculated by comparison of the spectra of D(+)-glucose anhydrous (Scharlab, S.L., Sentmenat, Barcelona, Spain), D(+)-xylose (Sigma-Aldrich Corp., St. Louis, MO, USA), D(+)-galactose (Panreac Química, S.L.U., Castellar del Vallès, Barcelona, Spain), L-rhamnose monohydrate (Sigma-Aldrich Corp., St. Louis, MO, USA), L(+)-arabinose (Sigma-Aldrich Corp., St. Louis, MO, USA), D(+)-mannose (Thermo Fisher Scientific, Geel, Belgium) and D(−)-fructose (Sigma-Aldrich Corp., St. Louis, MO, USA) commercial patterns. Data analysis was performed by Agilent ChemStation Revision B.04.03 SP1 software.

Total protein content was assessed following the procedure detailed by Bradford [35] with some changes. In short, Bradford reagent (0.4 mL) (Sigma-Aldrich Corp., St. Louis, MO, USA) was incorporated with each sample (1.6 mL). Then, the mixture was kept at room temperature for 5 min and the absorbance of this mixture was recorded at 595 nm. The same protocol was employed with bovine serum albumin (BSA) (Sigma-Aldrich Corp., St. Louis, MO, USA) as standard. The data were reported as mg BSA/ g raw material dry weight.

Total lipid content was colorimetrically determined following Kamal [36] analysis method with some modifications. Concisely, samples or standard (lauric acid) (500 µL) (Merck KGaA, Darmstadt,

Germany) were vortexed with *n*-hexane (1500 µL) (Panreac Química, S.L.U., Castellar del Vallès, Barcelona, Spain) for 1 min. After, the absorbance of the mixture was measured at 204 nm. The outcomes were expressed as g lauric acid/g raw material dry weight.

2.3.6. Color

The color measurements of all extracted liquid fractions were made within the CIEL*a*b* space using a colorimeter (CR-400, Konica Minolta, Japan). The obtained parameters were the lightness (whiteness, L* = 0, or brightness, L* = 100 degree), the red/green coordinate (degree of redness, a* > 0, or greenness, a* < 0) and yellow/blue coordinate (degree of yellowness, b* > 0, or blueness, b* < 0). Moreover, the corresponding magnitudes (Equations (1)–(3)) such as hue (h (°)), chroma (C*) and saturation (S*) were calculated as indicated below.

$$h(°) = \arctan\left(\frac{b^*}{a^*}\right). \tag{1}$$

$$C^* = \sqrt{(a^* + b^*)}. \tag{2}$$

$$S^* = \frac{C}{L^*}. \tag{3}$$

Additionally, the total color (ΔE*) and hue (ΔH*) differences (Equations (4) and (5)) were determined as follows.

$$\Delta E^* = \sqrt{(\Delta L^*)^2 + (\Delta a^*)^2 + (\Delta b^*)^2}. \tag{4}$$

$$\Delta H^* = \sqrt{(\Delta E^*)^2 - (\Delta L^*)^2 - (\Delta C^*)^2}. \tag{5}$$

2.3.7. Sun Protection Factor

Sun protection factor (SPF) was measured following Kaur and Saraf [37] methodology. The extract aliquots (20 µL) were homogeneously blended with 40% (*v/v*) ethanol solution (1980 µL) (Sigma-Aldrich Corp., St. Louis, Missouri, USA) and scanned between 290 and 320 nm. According to the Mansur et al. [38] procedure, the SPF values were determined using the following equation:

$$SPF = CF \, x \sum_{290}^{320} EE\,(\lambda) \times I\,(\lambda) \times Abs\,(\lambda). \tag{6}$$

where SPF corresponds to the spectrophotometric sun protection factor, CF represents a necessary correction factor (10), EE(λ) indicates the erythemal effect of the radiation with wavelength λ, I(λ) represents the solar intensity of the spectrum and Abs (λ) displays the spectrophotometric absorbance values at wavelength λ. The normalized product of EE(λ) and I(λ) utilized in the calculation of SPF was previously considered by Sayre et al. [39].

2.4. Formulation of the Creams

Sunscreen creams were elaborated at least in duplicate according to Balboa et al. [40] formulation. Briefly, components of oil phase (cream basis (O/W) (Derex, S.A., Rafelbunyol, Valencia, Spain) −18 g−, dimethicone 350 (Fagron Ibérica S.A.U., Terrassa, Barcelona, Spain) −6 g−, avocado oil (Fagron Ibérica S.A.U., Terrassa, Barcelona, Spain) −3 g−, sunscreen (Guinama, S.L.U., La Pobla de Vallbona, Valencia, Spain) −8 g−, micronized titanium dioxide (Fagron Ibérica S.A.U., Terrassa, Barcelona, Spain) −18 g− and fenonip XB (Fagron Ibérica S.A.U., Terrassa, Barcelona, Spain) −0.35 g−) were mixed and tempered at 70 ± 2 °C in a water bath. On the other hand, ingredients of water phase (distilled or thermal spring water −80 g−, carbomer 940 (Fagron Ibérica S.A.U., Terrassa, Barcelona, Spain) −1.5 g−, propyleneglycol (Guinama, S.L.U., La Pobla de Vallbona, Valencia, Spain) −6 g− and triethanolamine

(Fagron Ibérica S.A.U., Terrassa, Barcelona, Spain) −1.5 g−) were also blended and adjusted at 40 ± 2 °C. When the oil phase mixture was completely melted then it was transferred to the water bath at 40 ± 2 °C until its temperature adjustment. In this point, this phase was added at the water phase until their homogenization promoting a gel-like matrix. The studied extracts from the different raw materials by MHG at 100 W, butylhydroxytoluene (BHT) (Roig Farma, S.A., Terrasa, Barcelona, Spain) and (±)-α-tocopherol (Sigma-Aldrich Corp., St. Louis, MO, USA) commercial antioxidants or the different used water as controls (750 μL or mass equivalent weight), bergamot oil (450 μL) (Fagron Ibérica S.A.U., Terrassa, Barcelona, Spain) and tetramer cyclomethicone (3 mL) (Fagron Ibérica S.A.U., Terrassa, Barcelona, Spain) were mixed in the O/W emulsion at room temperature and thoroughly mixed. The final product was packed in flasks and amber glass vials and stored at refrigeration temperature until their analysis.

Water Features

Thermal spring waters were collected in plastic bottles in their points of emission (three different points of the Ourense region) and kept at 4 ± 2 °C in the laboratory within less the next 1 h after their gathering.

The electrical conductivity values of waters were determined by a conductivity meter (HI 9033, Hanna Instruments Ltd., Eden Way, UK) at room temperature. pH determination, total solid content level and color features of the different used waters were measured as indicated above.

2.5. Creams Characterization

Freshly elaborated sunscreen creams were examined so that their pH value, total solid content, SPF data, color characteristics, thiobarbituric acid reactive substances (TBARS) and viscous properties were determined. The effect of the elevated temperature and storage time on the stability of these samples was determined by means of accelerated oxidation at 50 ± 2 °C for 15 days. The pH and TBARS kinetics tests were done periodically on 1-day intervals whereas total solid content and color features assays were defined at the end of this process.

pH and color tests of the emulsions were realized as indicated above. The total solid content was determined using a portion of cream (0.5 g) under the above conditions. Sunscreen cream SPF values were also analyzed using a part of sunscreen cream (1 g) and 96% (*v/v*) ethanol (Sigma-Aldrich Corp., St. Louis, MO, USA) up to 25 mL following Dutra et al. [41] indications. The solution was stirred at 500 rpm for 5 min, treated by ultrasounds for 5 min and filtered. Absorbance data were read between 290 and 320 nm at 5 nm intervals. In this case, the correction factor was 20 and consequently, this value was considered in the mathematical calculation (Equation (6)). This value corresponded to a standard sunscreen formulation with a 5.46% solar filter equivalent to an SPF value of 3.18.

In order to evaluate the effectiveness of selected MHG extracts against lipid oxidation, thiobarbituric acid reactive substances (TBARS) was utilized following the assay of Scheffler et al. [42]. Shortly, samples or standard (malonaldehyde solution (20 nmol/mL) (Sigma-Aldrich Corp., St. Louis, MO, USA)) (0.8 g) were mixed with thiobarbituric acid-butylated hydroxytoluene (TBA-BHT) solution (1.6 mL) in tubes and shaken for 30 s using a vortex. This solution was prepared with 4,6-dihydroxy-2-mercaptopyrimide (15 g) (Thermo Fisher Scientific, Geel, Belgium), hydrochloric acid 12 M (1.76 mL) (Thermo Fisher Scientific, Geel, Belgium), distilled water (82.9 mL) and BHT (2%, *v/v*) (3 mL) (Roig Farma, S.A., Terrasa, Barcelona, Spain), previously treated with absolute ethanol (Sigma-Aldrich Corp., St. Louis, MO, USA). Afterward, test tubes were submitted to 95 ± 2 °C for 15 min in a water bath, cooled in an ice-tap-water bath for 10 min and centrifuged at 5000 rpm for 10 min. The absorbance of the supernatant was measured at 532 nm. The concentration of TBARS was expressed as nmol malonaldehyde/g sun cream dry weight.

The viscous profile of all formulated creams was determined. For this purpose, the apparent viscosity vs. shear rate ('so-called' steady-shear flow curves) was monitored at 25 °C on a controlled-stress rheometer (MCR 302, Paar Physica, Austria). The temperature was controlled by a Peltier system (± 0.01 °C). The selected measuring system was a sandblasted plate-plate (25 mm diameter, 1 mm gap) to avoid a possible slip of the samples. The edges of all creams placed on the plate-plate geometry were sealed with light paraffin oil to avoid water loss during experiments and were rested for 10 min to allow sample temperature and structural equilibration. In order to study the hysteresis effects, all steady-state shear measurements were made by decreasing and, subsequently, increasing shear rate following a logarithm. All viscous experiments were performed at least in triplicate.

2.6. Statistical Analysis

Experimental data were studied using one-factor analysis of variance, ANOVA (PASW Statistics v.22, IBM SPSS Statistics, New York, NY, USA). If the variance study indicated means differences, a Scheffé test was made to distinguish means with 95% confidence ($p < 0.05$).

3. Results and Discussion

3.1. Microwave Hydrodiffusion and Gravity (MHG)

Extraction time and temperature values by MHG from different samples are showed in Figure 1 in combination with the data of gathered extract volume. *Cytisus scoparius* flowers (Figure 1a) treated at 100 W provided 62.2 mL of total liquid phase for 150 min reaching a maximum temperature data of 92 °C. This raw material presented the typical microwave heating profile identified for other sources. Firstly, a latency period was passed after the first drop of liquid phase outside the microwave oven (around 8 min at 33 °C with a heating rate of 1.56 °C/min). Secondly, the next phase could be collected about 32% of the total volume for approximately 43 min. The following step exhibited a plateau region characterized by an average of 90 °C during ~52 min (heating rate of 0.14 °C/min) obtaining 48% of the total extract. At the last interval, the temperature decreased with a heating rate of 0.32 °C/min. This step (47 min) meant close to 20% of the total aqueous extract. Similar behavior was also displayed for *Pleurotus ostreatus* mushrooms and *Brassica rapa* leaves (Figure 1b,c). These samples, submitted to an MHG extraction process of 120 min, supplied 78 and 76.3 mL, respectively. The heating outline derived by the whole and divided into quarter acorns of *Quercus robur* L. did not describe the last phase since their moisture content (32.27 ± 3.79% wet basis, w.b.) was much lower than the other matrices. This variable is disclosed as a decisive element since in situ water of the tissue cells permits isolate the natural valuable compounds behaving like their transport media [43]. Processed acorns during 90 min exhibited roughly 30 mL of the liquid phase. The size reduction pretreatment of the samples favored slightly (9%) the extraction of their extracts since the influence of particle size is a key factor on the optimization of the extraction process as reported in the extraction of antioxidant compounds of other material such as tea or ginger by solid-liquid extraction [44].

The impact of MHG at 100 W on the liquid phase collected is displayed in Figure S1a. Samples provided a range of the initial volume of achieved water of 75–97%. The difference between the results was related to their moisture content. Note here that the moisture content of *C. scoparius* flowers, *P. ostreatus* mushrooms, and leaves of *B. rapa* was 82.79% ± 0.29% w.b., 89.43% ± 0.23% w.b. and 90.70% ± 0.71% w.b., respectively. These data agreed with those reported for other materials as the brown seaweed *Undaria pinnatifida* under similar extraction conditions [45]. The maximum total solid extraction yield (about 5.1 mg extract/g raw material dry weight) was obtained when mushroom samples were used (Figure S1b). Their structure tissue favored the removal of bioactive compounds in comparison with the other matrices. For example, their yield was ~5 times higher than the value derived by acorns samples. This pattern was also attributed to the energy consumption for the processing of these types of raw materials (Figure S1c) since acorns needed the lowest energy requirements (roughly 540 kJ). The energy consumption was elevated to approximately 900 kJ for to treat the wildflowers. The environmental impact defined as the quantity of carbon dioxide rejected into the atmosphere was in order to 120–200 g. This reflected the reduced burden of MHG extraction technology compared to conventional extraction methods [46].

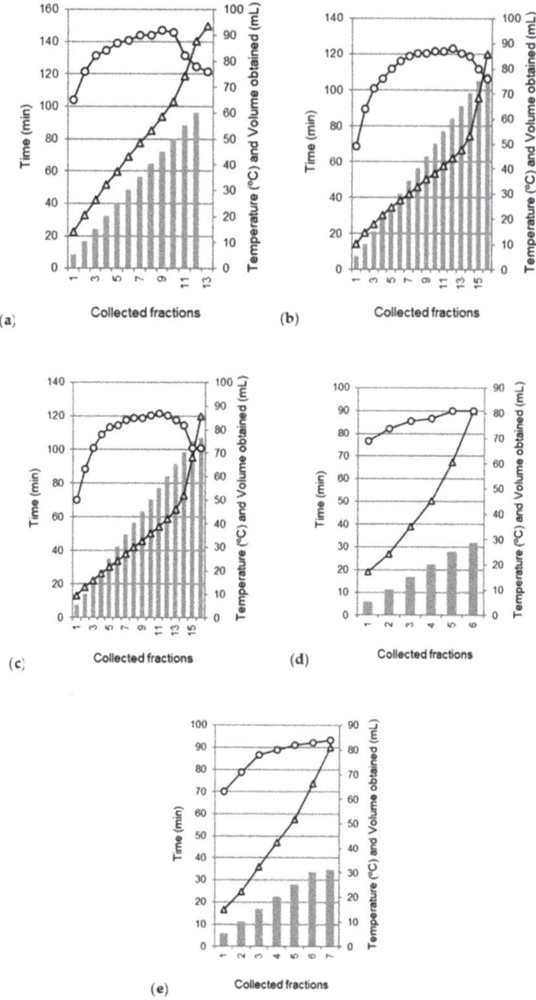

Figure 1. Collected volume (bars), vessel temperature (circles) and extraction time (triangles) during MHG treatment at 100 W of irradiation power of different raw materials: (**a**) defrosted *Cytisus scoparius* flowers, (**b**) defrosted fruiting bodies of *Pleurotus ostreatus* mushrooms, (**c**) defrosted *Brassica rapa* L. var. *rapa* leaves, (**d**) whole and (**e**) divided into quarters *Quercus robur* L. acorns. Note here that y-axes scales are different in each plot.

3.2. Antioxidant Properties of the Gathered Liquid Extracts

The total phenolic content and antioxidant potential of the extracts from the raw materials treated at 100 W by MHG technology are collected in Figure 2. Extracts from *C. scoparius* flowers showed the maximum total phenolic content, TEAC and FRAP values (around 0.23 mg GAE/g raw material dry weight, 0.68 mg Trolox eq/g raw material dry weight and 0.33 mg ascorbic acid/g raw material dry weight or 0.71 mg $FeSO_4 \times 7H_2O$/g raw material dry weight, respectively). Their total phenolic content agrees with the data previously found [12] for hydromethanolic and ethyl lactate extracts obtained by pressurized liquid extraction. On the other hand, *P. ostreatus* and *B. rapa* displayed similar results when both are compared. Specifically, their ferric reducing antioxidant power outcomes did not present significant differences among them (their average was about 0.12 mg ascorbic acid/g raw material dry weight or 0.28 mg $FeSO_4 \times 7H_2O$/g raw material dry weight). *B. rapa* showed the greatest antioxidant capacity with a noticeable concentration of 0.43 mg Trolox eq/g raw material dry weight and a total phenolic content higher than those reported by this sample [47]. Outcomes from the acorns of *Q. robur* indicated that samples divided into quarters provided a total phenolic content like the maximum data of wildflowers. This value was higher than reported by Rakić et al. [48] for similar raw materials submitted to methanolic extraction. This type of sample kept this tendency in the antioxidant capacity determined by the different free radical scavenging assays. The difference between the whole and the divided acorns was around 0.1 mg Trolox eq/g raw material dry weight for TEAC determination and 0.18 g ascorbic acid/g raw material dry weight or 0.39 mg $FeSO_4 \times 7H_2O$/g raw material dry weight for FRAP test. The DPPH percentage inhibition variation of the collected volume extracts from the two types of acorns was close to 57.8%. In this case, the divided raw material gave an effective concentration (EC_{50}, defined as the concentration in which 50% of DPPH radical was scavenged) of 0.05 g dry weight/100 g sample (equivalent to 0.67 mM of ascorbic acid).

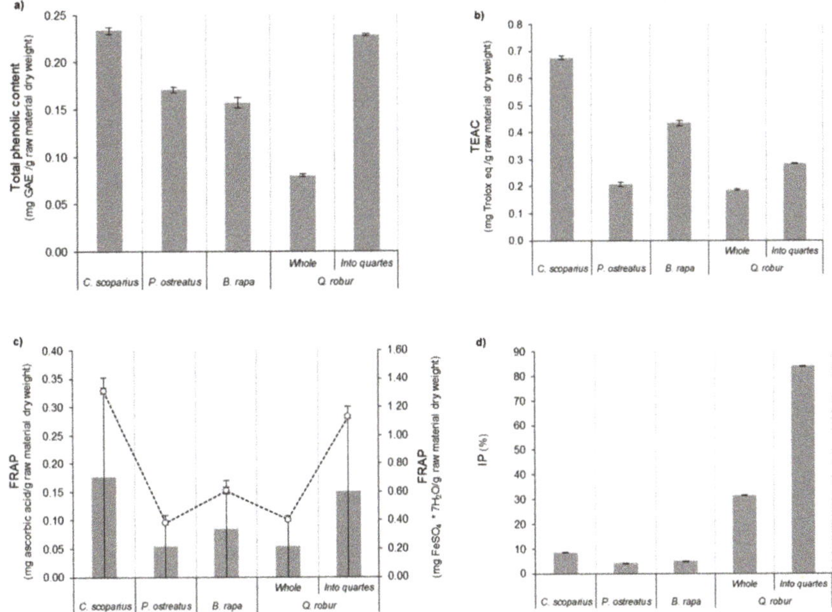

Figure 2. Total phenolic content (**a**) and antioxidant capacity, expressed as TEAC value (**b**), FRAP assay (**c**) as ascorbic acid equivalents (circles) and as ferric sulphate equivalents (bars) and inhibition percentage (IP) of DPPH radical (**d**), from the aqueous collected of the different took in consideration raw materials at 100 W by MHG extraction.

Carotenoids are pigments integrated by lipid-soluble tetraterpenoid compounds. Their structure can affect their antioxidant potential scavenging efficiently O_2 and peroxyl radicals [49,50]. The total carotenoid content of the liquid extracts is summarized in Figure 3a. Wildflowers provided the maximum concentration (3.1 µg β-carotene/g raw material dry weight). This value could be attributed to its antioxidant profile since it was more elevated as stated above. In comparison with mushroom data, their concentration was around three times superior-this relation differed to their antioxidant profile-. The samples of acorn presented the lowest total carotenoid content: the samples divided in quarters allowed a result of about 32.3% higher than the whole nuts. In the current study, the raw materials used by MHG procedure were also submitted to a solid-liquid extraction (SLE) with 70–90% (*v/v*) ethanol solution as a solvent for evaluation purposes. The resultant total carotenoid content appears in Figure 3b. In contrast with the MHG outcomes, *B. rapa* samples supplied the maximum concentration registered of total carotenoids content (approximately 3.1 µg *β*-carotene/g raw material dry weight). This value was the highest outcome described by *Brassica rapa* ssp. *chinensis* sprouts [51]. The opposite trend was followed by the remaining samples since these showed issues lower than the reported by MHG procedure (with a difference in the range between 0.34 and 2.73 units).

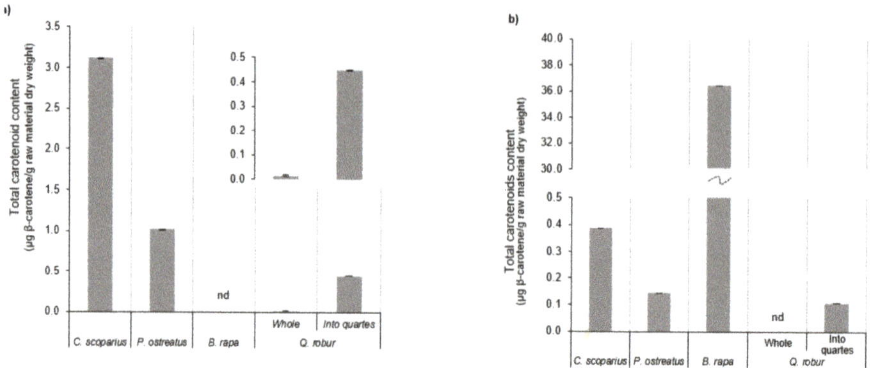

Figure 3. Total carotenoids content from the collected aqueous extracts of the different took in consideration raw materials at 100 W by MHG extraction (**a**) and ethanol extracts collected by SLE (**b**) of the evaluated feedstocks. nd: no detected.

The antioxidant profile of the ethanolic extracts from the different employed resources by SLE is exhibited in Figure 4. Despite the fact that the total phenolic content of the *C. scoparius* flowers was limited (close to 16 mg GAE/g raw material dry weight), their antioxidant potential was defined by considerable TEAC and DPPH values (around 30 mg Trolox eq/g raw material dry weight and 71.0% inhibition percentage, respectively). Their EC_{50} was 0.95 ± 0.01 g/100 g dry weight (equivalent to 0.60 mM of ascorbic acid). Note here that these wildflowers showed the maximum total solid extraction yield (close to 5.8 mg extract/g flower dry weight). The antioxidant data from *P. ostreatus* samples did not stand out when compared with the other raw materials, although their concentrations were higher than the values obtained by MHG. This agrees with the tendency described by the other sources. On the other hand, *B. rapa* allowed the largest TEAC parameter (42.2 mg Trolox eq/g raw material dry weight) whereas their FRAP and DPPH magnitudes were intermediate (16.5 mg ascorbic acid/g raw material dry weight or 42.8 mg $FeSO_4 \times 7H_2O$/g raw material dry weight and 36.0% inhibition percentage, respectively). Even though, *Q. robur* acorns divided into quarters had not an important TEAC concentration, this sample provided the maximum data of total phenolic content (136.9 mg GAE/g raw material dry weight), FRAP values (36.9 mg ascorbic acid/g raw material dry weight or 94.1 mg $FeSO_4 \times 7H_2O$/g raw material dry weight) and DPPH inhibition percentage (93.0%) with significant difference in comparison with the remaining resources. Overall, it is necessary to emphasize that a strong positive correlation was displayed between total phenolic content and ascorbic acid and iron (II) sulphate heptahydrate results (r = 0.950 and r = 0.948, respectively).

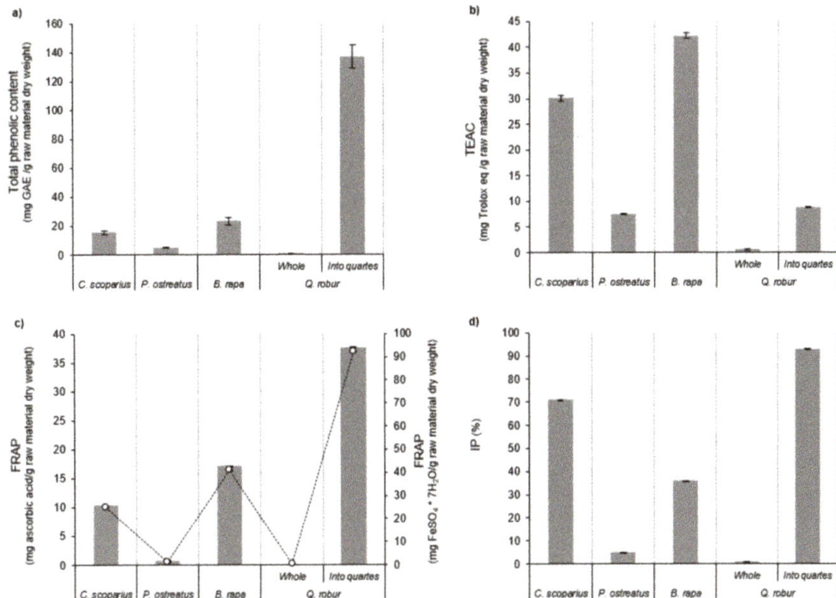

Figure 4. Total phenolic content (**a**) and antioxidant capacity, expressed as TEAC value (**b**), FRAP assay (**c**) as ascorbic acid equivalents (circles) and as ferric sulphate equivalents (bars) and inhibition percentage (IP) of DPPH radical (**d**), from the ethanol extracts by SLE of the different raw materials.

3.3. Proximate Chemical Profile

The collected aqueous extracts by the MHG method and ethanol extracts derived by SLE extraction were submitted to macronutrients analysis. Their total protein and total lipid contents are presented in Figure 5. *B. rapa* and *P. ostreatus* samples showed the highest total protein content by MHG and SLE extraction procedures, respectively. These data varied between approximately 2.1 and 72.0 mg BSA/g raw material dry weight. Generally, the tendency of SLE extracts corresponded with the behavior defined by their total carotenoid content. This trend could be defined by the chemical structure of the vegetable pigments since they are linked to some chlorophyll-carotenoid proteins [47]. In relation to total lipid content of the gathered extracts, *Q. robur* divided hard fruits supplied the highest outcome by MHG (783.9 g lauric acid/g raw material dry weight) with significant differences as compared with the other feedstocks. The ethanolic extracts provided the lowest total lipid content issues. The obtained data showed that ethanol solvent at $40 \pm 2\ °C$ did not permit recovering lipid compounds from the different samples. These reduced lipid yields could be improved using a multiple-batch sequence of organic solvents at greatest temperature ranges able to removal apolar compounds as occurred with other raw materials as rice bran or coffee grounds [52,53]. Further studies should be interesting to increase the lipid extraction from the used raw materials.

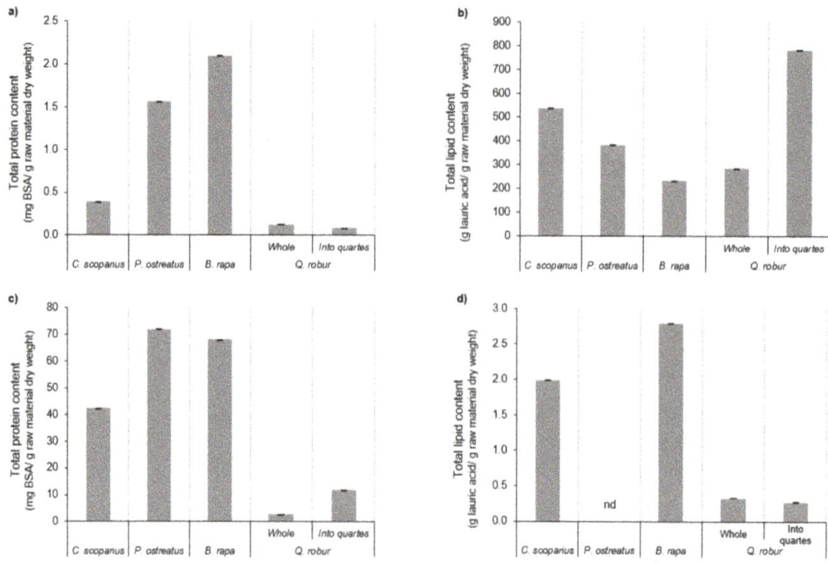

Figure 5. Total protein content (**a,c**) and total lipid content (**b,d**) from the collected aqueous extracts of the different took in consideration raw materials at 100 W by MHG extraction and from the SLE ethanolic extracts, respectively, from different raw materials. nd: no detected.

The monosaccharide and oligosaccharide composition of the extracts from MHG and SLE methodologies are summarized in Table 1. The sample that presented the highest carbohydrate concentration was *B. rapa* followed by *P. ostreatus* and *C. scoparius*. Overall, SLE could extract higher quantities of carbohydrate compounds than the MHG procedure. Mannose was the main monosaccharide identified in all samples. In all cases, the corresponding oligosaccharide was also the major compound. Turnip sample displayed a higher concentration of oligosaccharides than monosaccharides (about double times). This behavior was in accordance with the literature data from *Brassica oleracea* leaves [54]. Different studies described that mannose is present in the structure of the functional protein as ligands or signaling molecules [55,56]. A similar tendency was observed in this monosaccharide and the total protein content of SLE extracts suggested a relation between both components. The extraction capacity of the MHG method varied between around 0.02% and 95.65% as well as between 0.05% and 0.44% of the total recovery monosaccharides and O-mannose compounds obtained by SLE. *Q. robur* acorn exhibited the lowest data. It is worth noting that MHG whole acorns aqueous extracts displayed an opposite trend in their xylose, rhamnose and mannose monosaccharides concentration. This sample divided in quarters also shared this tendency for the values of their rhamnose and O-glucose oligosaccharide compounds.

Table 1. Influence of the extraction at 100 W by MHG and solid liquid extraction from the different raw materials on the monosaccharide and oligosaccharide composition (mg/g raw material dry weight) of the liquid phase collected.

Composition (mg/g Raw Material Dry Weight)		Extraction Technique	Raw Materials				
			C. scoparius	*P. ostreatus*	*B. rapa*	*Q. robur*	
						Whole	Into Quarters
Monosaccharides	Glucose	MHG	0.34 ± 0.00 [j]	0.68 ± 0.00 [i]	0.32 ± 0.00 [m]	0.01 ± 0.00 [g]	0.01 ± 0.00 [k]
		SLE	10.60 ± 0.00 [e]	2.89 ± 0.00 [g]	9.31 ± 0.01 [f]	0.01 ± 0.00 [g]	0.10 ± 0.00 [g]
	Xylose	MHG	0.08 ± 0.01 [l]	0.17 ± 0.00 [l]	0.17 ± 0.00 [n]	0.03 ± 0.00 [e]	0.01 ± 0.00 [l]
		SLE	2.36 ± 0.01 [h]	0.47 ± 0.01 [j]	1.03 ± 0.01 [i]	0.03 ± 0.01 [e]	0.03 ± 0.00 [i]
	Galactose	MHG	0.04 ± 0.00 [ll]	0.07 ± 0.00 [m]	0.12 ± 0.01 [ñ]	0.02 ± 0.01 [f]	0.01 ± 0.00 [k]
		SLE	0.37 ± 0.00 [j]	0.34 ± 0.00 [k]	0.62 ± 0.01 [k]	0.02 ± 0.01 [f]	0.03 ± 0.00 [j]
	Rhamnose	MHG	0.01 ± 0.00 [m]	0.13 ± 0.00 [ll]	0.05 ± 0.00 [p]	0.01 ± 0.01 [g]	0.03 ± 0.00 [i]
		SLE	0.13 ± 0.00 [k]	0.16 ± 0.01 [l]	0.39 ± 0.01 [ll]	0.01 ± 0.00 [g]	0.01 ± 0.00 [k]
	Arabinose	MHG	0.45 ± 0.00 [i]	nd	0.44 ± 0.00 [l]	nd	0.07 ± 0.00 [h]
		SLE	nd	nd	nd	nd	nd
	Mannose	MHG	0.37 ± 0.00 [j]	6.11 ± 0.03 [f]	0.85 ± 0.00 [j]	97.03 ± 2.78 [b]	0.78 ± 0.00 [e]
		SLE	1537.0 ± 2.4 [b]	1551.3 ± 2.8 [b]	2253.5 ± 0.8 [b]	96.44 ± 0.80 [b]	107.20 ± 0.37 [b]
	Fructose	MHG	nd	1.28 ± 0.01 [h]	0.61 ± 0.01 [k]	0.47 ± 0.31 [d]	0.03 ± 0.00 [j]
		SLE	5.14 ± 0.48 [f]	10.21 ± 0.39 [e]	8.32 ± 0.01 [g]	0.76 ± 0.65 [c]	0.41 ± 0.02 [f]
Oligosaccharides	O-glucose	MHG	nd	nd	0.07 ± 0.00 [o]	0.01 ± 0.00 [h]	0.03 ± 0.00 [j]
		SLE	nd	0.72 ± 0.00 [i]	nd	nd	0.02 ± 0.05 [k]
	O-xylose	MHG	nd	nd	nd	nd	nd
		SLE	43.96 ± 0.15 [d]	13.47 ± 0.02 [d]	18.17 ± 0.10 [e]	nd	0.43 ± 0.17 [f]
	O-galactose	MHG	nd	nd	nd	nd	nd
		SLE	nd	nd	nd	nd	nd
	O-rhamnose	MHG	54.33 ± 0.31 [c]	572.3 ± 0.0 [c]	599.2 ± 4.4 [d]	nd	7.16 ± 2.24 [c]
		SLE	nd	nd	nd	nd	nd
	O-arabinose	MHG	3.42 ± 0.05 [g]	nd	3.62 ± 0.05 [h]	nd	0.01 ± 0.05 [k,l]
		SLE	nd	nd	nd	nd	nd
	O-mannose	MHG	nd	2.10 ± 0.39 [g]	nd	nd	1.19 ± 0.22 [d]
		SLE	2880.5 ± 0.7 [a]	4492.9 ± 0.6 [a]	4059.0 ± 8.2 [a]	166.5 ± 2.3 [a]	271.2 ± 58.3 [a]
	O-fructose	MHG	nd	nd	706.03 ± 1.62 [c]	nd	nd
		SLE	nd	nd	nd	nd	nd

Data are given as mean ± standard deviation. Data values in a column with different superscript letters are statically different ($p \leq 0.05$). nd: no detected.

3.4. Color Characteristics of the Extracts

Colorimetric coordinates from the CIEL*a*b* method were measured directly to the extracts from the different samples by MHG at 100 W and SLE extractions. Considering the above indices, the color magnitudes were calculated according to the equations stated above. These data are reported in Table 2. In general, lightness, a* coordinates and the hue angle of the ethanolic extracts by SLE showed lower values than the aqueous extracts by MHG. The opposite tendency was observed by the b* coordinate and Chroma and saturation parameters. Significant differences were identified between the tested extraction procedures for each resource. Particularly, the hue angle of ethanolic extract from acorns divided into quarters disclosed higher differences between the two extraction systems (64.9°). The variation of the color features of the gathered extracts was also clear when the total color difference (ΔE^*) of the samples from MHG was compared with the parameters derived by the SLE technique, according to the classification of Adekunte et al. [57]. This change was classified as a small difference ($\Delta E^* <$ 1.5) for whole acorns and different ($1.5 < \Delta E^* < 3.0$) for divided nuts and mushrooms. These results could indicate that these extraction methodologies provided similar color extracts from these samples. However, the application of MHG and SLE processes from wildflowers and turnip leaves could practice an effect more dissimilar under these plant matrices since their value of total color difference was very different ($\Delta E^* > 3.0$). Concerning the hue difference (ΔH^*) carried out with a trend like the total color difference in order that the *Q. robur* acorns and *B. rapa* samples presented the lowest and the highest hue difference values, respectively. It is necessary to note that a strong negative correlation between L* and a* coordinates and total carotenoid content was noticeable (r = −0.912 and r = −0.982, respectively). Conversely, a positive correlation with this concentration and the b* coordinate, C* and S* magnitudes was identified (r = 0.971–0.987). Based on these Pearson's correlation coefficients, the relationship of total carotenoid content with parameters from the CIELab color space will allow estimate rapidly this pigment concentration of extracts by MHG and SLE technologies with a non-destructive method. Similar associations were established for several vegetable raw materials [58,59].

3.5. Utilization of MHG Extracts on Cosmetic Formulations

Natural extracts obtained from underused and by-products raw materials can act as functional additives in cosmetic products contributing to healthy properties. These sustainable sources are not expensive and appear in abundance. The environmental, social and economic impact of the production of these extracts is also evaluated by cosmetic industries since nowadays the preservation of the environment and a positive influence on society sphere are objectives to be achieved. Currently, there is a growing interest in the incorporation of these products in cosmetic formulations in response to the increasing demand by consumers [40,60]. The potential cosmeceutical use of the *C. scoparius*, *P. ostreatus*, *B. rapa* and *Q. robur* acorns divided into quarters MHG extracts was researched by the incorporation of this aqueous extracts as an additive ingredient at sunscreen cream formulations. The added extracts displayed an approximately pH value on the range from 3.1 to 5.3. Their SPF data varied between approximately 0.04 and 0.21. It is remarkable that flower extract provided the highest SPF issue whereas their pH was the lowest. Note here that the SPF value of the elaborated creams was around 5.8. Concerning the enzymatic inhibition assays (data not shown), the *Q. robur* sample provided the highest anti-elastase capacity with an effective concentration (EC_{50}) of 606 mg extract/L (equivalent to 2.3 mg epigallocatechin gallate/L) and an EC_{50} tyrosinase inhibitory capacity of 496 mg extract/L (equivalent to 93 mg kojic acid/L). These preliminary results suggested that this extract may be helpful in preventing the loss of skin sagging and elasticity and the pigmentation damage [61].

Table 2. Colorimetric features by CIEL*a*b* system from the aqueous phase collected at 100 W by MHG extraction and solid liquid extraction of the different raw materials.

Raw Materials		Extraction Technique	Coordinates			Magnitudes		
			Lightness (L*)	a*	b*	Hue Angle (h°)	Chroma (C*)	Saturation (S*)
C. scoparius		MHG	90.10 ± 0.03 [a]	0.65 ± 0.01 [b]	−1.44 ± 0.02 [f]	114.34 ± 0.12 [d]	1.58 ± 0.01 [e]	0.02 ± 0.00 [b]
		SLE	87.50 ± 0.03 [c]	−1.06 ± 0.04 [d]	3.09 ± 0.06 [b]	108.14 ± 0.11 [g]	3.27 ± 0.08 [b]	0.04 ± 0.01 [b]
P. ostreatus		MHG	90.07 ± 0.02 [a]	0.71 ± 0.02 [a]	−1.59 ± 0.02 [g]	114.11 ± 0.10 [d]	1.74 ± 0.01 [c]	0.02 ± 0.01 [b]
		SLE	87.67 ± 0.05 [c]	0.23 ± 0.04 [c]	−0.67 ± 0.02 [d]	109.14 ± 0.13 [f]	0.71 ± 0.02 [i]	0.01 ± 0.00 [b]
B. rapa		MHG	90.04 ± 0.02 [a]	0.71 ± 0.02 [a]	−1.65 ± 0.01 [g]	113.24 ± 0.10 [e]	1.80 ± 0.01 [c]	0.02 ± 0.01 [b]
		SLE	83.99 ± 0.07 [d]	−6.45 ± 0.05 [e]	13.38 ± 0.07 [a]	115.77 ± 0.16 [c]	14.85 ± 0.15 [a]	0.18 ± 0.03 [a]
Q. robur	Whole	MHG	90.15 ± 0.02 [a]	0.66 ± 0.02 [a,b]	−1.23 ± 0.01 [e]	118.22 ± 0.10 [b]	1.40 ± 0.02 [f]	0.02 ± 0.01 [b]
		SLE	90.31 ± 0.04 [a]	0.68 ± 0.01 [a]	−1.50 ± 0.07 [f,g]	114.36 ± 0.14 [d]	1.65 ± 0.01 [d]	0.02 ± 0.00 [b]
	Into quarters	MHG	89.25 ± 0.03 [b]	0.62 ± 0.04 [b]	−0.68 ± 0.04 [d]	132.06 ± 0.13 [a]	0.92 ± 0.01 [h]	0.01 ± 0.00 [b]
		SLE	89.64 ± 0.05 [b]	0.29 ± 0.03 [c]	1.27 ± 0.08 [c]	67.21 ± 0.15 [h]	1.33 ± 0.01 [g]	0.01 ± 0.00 [b]

	Total Color Difference (ΔE*)			Hue Difference (ΔH*)		
MHG / SLE		*C. scoparius*	5.50		*C. scoparius*	4.54
		P. ostreatus	2.61		*P. ostreatus*	0.13
		B. rapa	17.71		*B. rapa*	10.34
	Q. robur	Whole	0.31	*Q. robur*	Whole	0.10
		Into quarters	2.02		Into quarters	1.93

Data are given as mean ± standard deviation. Data values in a column with different superscript letters are statically different ($p \leq 0.05$).

These O/W emulsions were formulated with three different types of thermal spring waters and with distilled water for comparative purposes. Their pH values ranged from 5.46 to 7.57. These parameters were in accordance with the values of the pH reported by Delgado-Outeiriño et al. [62] for different samples from thermal spring waters of the same region. The electrical conductivity of the thermal water samples varied on the range of 1260–3200 µS so that the thermal spring water 1 displayed the highest value in accordance with their total solid content (0.45 mg dry residue/mL). The remaining samples exhibited an average of ~0.13 mg dry residue/mL. No significant differences were provided between these samples (data not shown). This behavior was also provided in the color magnitudes of the water samples. Their hue angle was about 136.4° whereas their Chroma and saturation were thereabouts 5.6 and 0.06, respectively.

Figure S2 discloses the total solid content of the sunscreen creams freshly processed. Results obtained showed that no significant differences were detected between the different types of water and the different extracts or commercial antioxidants. This parameter varied in a limited interval of approximately 0.31 and 0.35 g dry residue/g sun cream, respectively. These creams submitted at the accelerated oxidation at 50 ± 2 °C for 15 days displayed a total solid content lower, with minimum and maximum values of 0.328 and 0.330 g dry residue/g sun cream, respectively. In relation to the color features of the creams made at the initial time (data not shown), the measurements exhibited similar issues with small differences ($\Delta E^* < 1.5$) for their total color parameter. The highest data was provided for *P. ostreatus* and/with thermal spring water 2 cream ($\Delta E^* = 0.84$). This sample also exhibited the highest value of this same parameter with an outcome of $\Delta E^* = 2.00$ after 15 days of stability testing. In this case, the oxidation chemical reactions that were produced derived into an increase of the total color difference of the tested cosmetics, since different differences ($1.5 < \Delta E^* < 3.0$) were registered in thermal spring water 2 cream with all added antioxidant ingredients and *Q. robur* sun cream with thermal spring water 3.

Kinetics of pH values for 15 days of the accelerated assays were realized and representative values were gathered in Figure S3. Overall, the pH of the creams varied between 5.3 and 6.3. Distilled water was the type of water that offered the pH data with more fluctuation (ranged from 5.3 to 5.9) in contraposition with water from thermal spring water 3 (with a variation of about 0.2 points). These data were compatible with the normal skin pH [63]. Thermal water spring 1 control cream supplied data slightly higher than added extracts and antioxidant creams made with this same type of water. This behavior could be associated with the acid pH of the added extracts and the BHT and (±)-α-tocopherol features in combination with characteristics of the different used ingredients in this model cosmetic. This observation was not reflected in the evolution of the TBARS values of these same samples (Figure S4). In this case, no significant differences were recorded for the different studied emulsions elaborated with each type of water. Generally, a plateau region was described by the malonaldehyde values of the different sun creams. The data of this marker of lipid peroxidation process ranged from 11.8 to 26.7 nmol malonaldehyde/g sun cream dry weight. Several scientific studies reported a similar trend for the TBARS index in O/W emulsions [64,65].

Figure 6 summarizes the effect of the three tested thermal waters and distilled water in the flow behavior at 25 °C of creams incorporated with different natural extracts (*C. scoparius*, *P. ostreatus*, *B. rapa*, *Q. robur*). For all tested systems, the apparent viscosity showed shear-thinning behavior, dropping around four decades with an increasing shear rate. At the fixed shear rate, the presence of thermal spring water 3 promoted the decrease of the apparent viscosity, especially for creams formulated with *P. ostreatus* and *B. rapa* (Figure 6b,c). This is an indicator of the relevance of studying different extracts because the rheological properties greatly depend on the analyzed extract and its specifically composition. This suggests that water from the thermal spring 3 can provide sunscreen creams that will be easily applicable in the skin [66,67]. Concerning magnitudes of the apparent viscosity of the creams formulated with different extracts, results suggest that those that presented higher bioactive compounds also exhibited lower apparent viscosity values. This agrees with the achievements provided for other different functional matrices [68–70]. Concerning BHT and (±)-α-tocopherol creams (Figure 6e,f), similar profiles and magnitudes were identified, whereas those made in the water control (Figure 6g) displayed slightly higher values. Again, no hysteresis loops were observed in tested sunscreen creams.

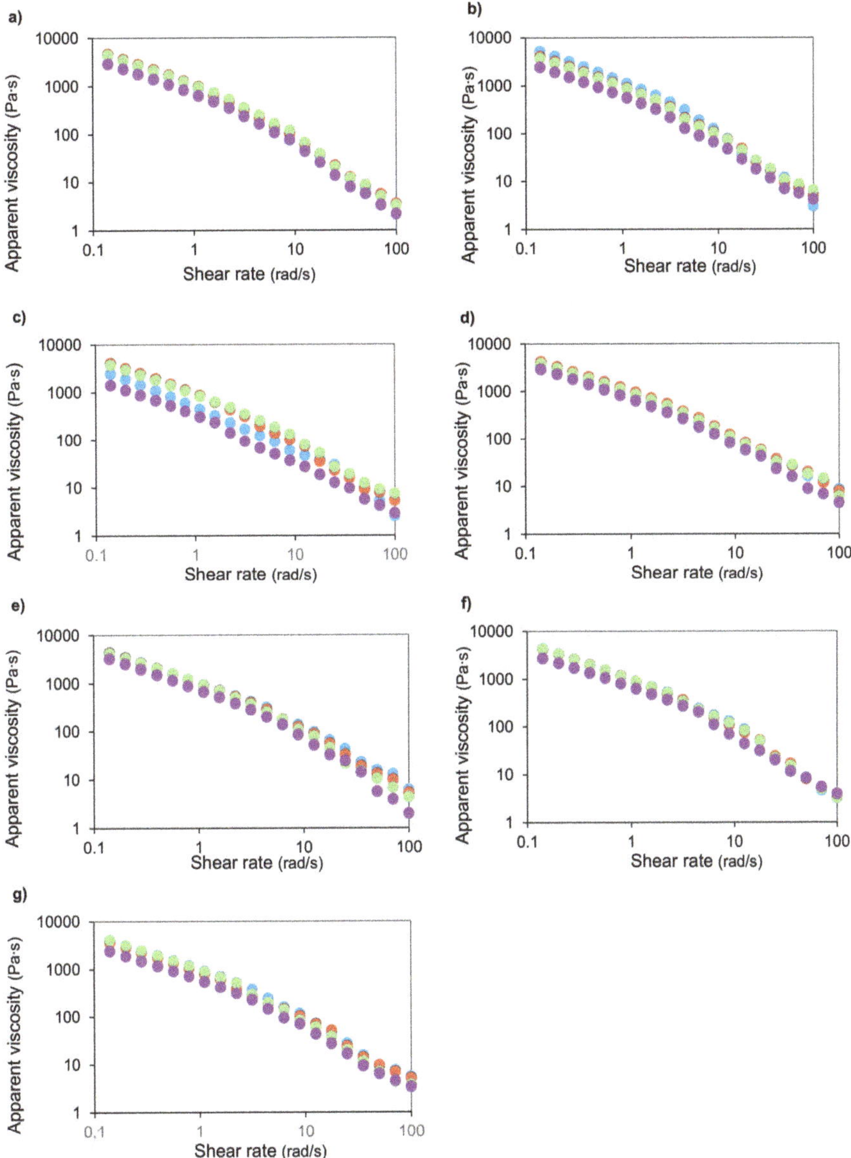

Figure 6. Viscous behavior of the sun cream formulated with selected MHG extracts from (**a**) *C. scoparius* flowers, (**b**) *P. ostreatus* mushrooms, (**c**) *B. rapa* leaves and (**d**) *Q. robur* acorns and (**e**) BHT, (**f**) (±)-α-tocopherol and (**g**) water control. Color code: distilled water (blue), thermal spring water 1 (red), thermal spring water 2 (green) and thermal spring water 3 (purple).

4. Conclusions

Emerging MHG technology was found to be appropriate to produce extracts with antioxidant properties. The highest total solid extraction yield was obtained from *P. ostreatus* mushroom whereas the lowest energy requirements were identified for *Q. robur* acorns. This last sample divided into quarters also displayed the highest DPPH data. *C. scoparius* flowers provided the extracts with the highest total phenolic content, TEAC, and FRAP values and total carotenoid content. *B. rapa* extracts from MHG was not relevant in contraposition with their ethanolic extracts that disclosed relevant bioactive properties in combination with acorns divided into quarters. These samples showed the most noticeable total lipid content by SLE and MHG, respectively. *B. rapa* supplied the highest total protein content and carbohydrate concentration by MHG extraction systems. The color features and total carotenoid content of the studied extracts presented a strong correlation. In relation to the cosmetic application of the MHG extracts, *Q. robur* acorns divided into quarters registered the highest anti-elastase and anti-tyrosinase capacities. The use of these extracts into sunscreen creams with three different thermal spring waters disclosed that their chemical and bioactive features were like those with added commercial antioxidants. The combination of *P. ostreatus* and *B. rapa* extracts with thermal spring water 3 offered the most suitable mechanical properties favoring the skin cream application. Further studies need to focus on improving the rheological properties at different temperatures over time in combination with the sensorial analysis of the formulated cosmetic emulsions. Finally, MHG lab-scale studies should be applied to pilot scale, following the suitable results previously found for the MHG technique applied to pilot-scale solvent-free microwave extraction of polyphenols from *Lettuce sativa* [71].

Supplementary Materials: The following are available online. Figure S1: Effect of the microwave irradiation power during microwave hydrodiffusion and gravity (MHG) extraction at 100 W (a) aqueous phase collected and (b) extraction yield of total solid of aqueous phase recovered of the different supplied raw materials. Values of the required energy consumption required are also summarized for comparative purposes; Figure S2: Total solid content of the sun cream formulated with distilled or thermal waters and with selected MHG extracts from defrosted *C. scoparius* flowers (grey bars), defrosted fruiting bodies of *P. ostreatus* mushrooms (bars with vertical lines), defrosted *B. rapa* leaves (texture bars) and quarters of *Q. robur* acorns (zigzag bars). For comparative purpose, data of sun cream formulated with BHT (bars with horizontal lines), (±)-α-tocopherol (dotted bars) and water control (squared bars) are also showed; Figure S3: Evolution of pH values during accelerated oxidation determination of the sun cream formulated with distilled or thermal waters and with selected MHG extracts from defrosted *C. scoparius* flowers (dark grey bars), defrosted fruiting bodies of *P. ostreatus* mushrooms (light grey bars), defrosted *B. rapa* leaves (grey bars) and quarters of *Q. robur* acorns (bars with vertical lines) and BHT (bars with horizontal lines), (±)-α-tocopherol (dotted bars) and water control (squared bars) with (a) distilled water, (b) thermal spring water 1, (c) thermal spring water 2 and (d) thermal spring water 3; Figure S4: Evolution of thiobarbituric acid reactive substances (TBARS) during accelerated oxidation determination of the sun cream formulated with distilled or thermal waters and with selected MHG extracts from defrosted *C. scoparius* flowers (dark grey bars), defrosted fruiting bodies of *P. ostreatus* mushrooms (light grey bars), defrosted *B. rapa* leaves (grey bars) and quarters of *Q. robur* acorns (bars with vertical lines) and BHT (bars with horizontal lines), (±)-α-tocopherol (dotted bars) and water control (squared bars) with (a) distilled water, (b) thermal spring water 1, (c) thermal spring water 2 and (d) thermal spring water 3.

Author Contributions: Conceptualization, E.F., H.D. and M.D.T.; data curation, L.L.-H. and M.D.T.; writing—original draft preparation, L.L.-H. and M.D.T.; writing—review and editing, E.F., H.D. and M.D.T.; funding acquisition, L.L.-H., H.D. and M.D.T. All authors have read and agreed to the published version of the manuscript.

Funding: This research was funded by Ministry of Economy and Competitiveness of Spain (CTM2015-68503-R). L. López-Hortas thanks the Xunta de Galicia for her pre-doctoral grant (2014/2020 European Social Fund). M.D. Torres acknowledges the Ministry of Economy, Industry and Competitiveness for her post-doctoral grant (IJCI-2016-27535).

Conflicts of Interest: The authors declare no conflict of interest.

References

1. Kozłowska, J.; Prus, W.; Stachowiak, N. Microparticles based on natural and synthetic polymers for cosmetic applications. *Int. J. Boil. Macromol.* **2019**, *129*, 952–956. [CrossRef] [PubMed]
2. Rafiee, Z.; Nejatian, M.; Daeihamed, M.; Jafari, S.M. Application of curcumin-loaded nanocarriers for food, drug and cosmetic purposes. *Trends Food Sci. Technol.* **2019**, *88*, 445–458. [CrossRef]

3. An, J.; Lee, I.; Yi, Y. The Thermal Effects of Water Immersion on Health Outcomes: An Integrative Review. *Int. J. Environ. Res. Public Health* **2019**, *16*, 1280. [CrossRef] [PubMed]

4. Matsumoto, S. Evaluation of the Role of Balneotherapy in Rehabilitation Medicine. *J. Nippon. Med Sch.* **2018**, *85*, 196–203. [CrossRef]

5. Ferreira, E.B.; Vasques, C.I.; Gadia, R.; Chan, R.J.; Guerra, E.N.S.; Mezzomo, L.A.; De Luca Canto, G.; dos Reis, P.E.D. Topical interventions to prevent acute radiation dermatitis in head and neck cancer patients: A systematic review. *Support. Care Cancer* **2017**, *25*, 1001–1011. [CrossRef]

6. Maarouf, M.; Saberian, C.; Lio, P.A.; Shi, V.Y. Head-and-neck dermatitis: Diagnostic difficulties and management pearls. *Pediatr. Dermatol.* **2018**, *35*, 748–753. [CrossRef]

7. Zeichner, J.; Seite, S. From probiotic to prebiotic using thermal spring water. *J. Drugs Dermatol.* **2018**, *17*, 657–662.

8. Afonso, T.; Moresco, R.; Uarrota, V.G.; Navarro, B.B.; Nunes, E.D.C.; Maraschin, M.; Rocha, M. UV-Vis and CIELAB Based Chemometric Characterization of Manihot esculenta Carotenoid Contents. *J. Integr. Bioinform.* **2017**, *14*, 20170056. [CrossRef]

9. Morone, J.; Alfeus, A.; Vasconcelos, V.; Martins, R. Revealing the potential of cyanobacteria in cosmetics and cosmeceuticals—A new bioactive approach. *Algal Res.* **2019**, *41*, 101541. [CrossRef]

10. Wongwad, E.; Pingyod, C.; Saesong, T.; Waranuch, N.; Wisuitiprot, W.; Sritularak, B.; Temkitthawon, P.; Ingkaninan, K. Assessment of the bioactive components, antioxidant, antiglycation and anti-inflammatory properties of Aquilaria crassna Pierre ex Lecomte leaves. *Ind. Crops Prod.* **2019**, *138*, 111448. [CrossRef]

11. González, N.; Ribeiro, D.; Fernandes, E.; Nogueira, D.R.; Conde, E.; Moure, A.; Vinardell, M.P.; Mitjans, M.; Domínguez, H. Potential use of Cytisus scoparius extracts in topical applications for skin protection against oxidative damage. *J. Photochem. Photobiol. B Boil.* **2013**, *125*, 83–89. [CrossRef]

12. Lores, M.; Pajaro, M.; Alvarez-Casas, M.; Domínguez, J.; Garcia-Jares, C. Use of ethyl lactate to extract bioactive compounds from Cytisus scoparius: Comparison of pressurized liquid extraction and medium scale ambient temperature systems. *Talanta* **2015**, *140*, 134–142. [CrossRef]

13. Taofiq, O.; Heleno, S.A.; Calhelha, R.C.; Fernandes, I.P.; Alves, M.J.; Barros, L.; González-Paramás, A.M.; Ferreira, I.C.; Barreiro, M.F. Mushroom-based cosmeceutical ingredients: Microencapsulation and in vitro release profile. *Ind. Crops Prod.* **2018**, *124*, 44–52. [CrossRef]

14. Taofiq, O.; Rodrigues, F.; Barros, L.; Barreiro, M.F.; Ferreira, I.C.; Oliveira, M.B.P. Mushroom ethanolic extracts as cosmeceuticals ingredients: Safety and ex vivo skin permeation studies. *Food Chem. Toxicol.* **2019**, *127*, 228–236. [CrossRef]

15. Sena, L.M.; Zappelli, C.; Apone, F.; Barbulova, A.; Tito, A.; Leone, A.; Olivoero, T.; Ferracane, R.; Fogliano, V.; Colucci, G. Brassica rapa hairy root extracts promote skin depigmentation by modulating melanin production and distribution. *J. Cosmet. Dermatol.* **2018**, *17*, 246–257. [CrossRef]

16. Subramanian, V.; Sahithya, D. Preliminary Screening of Selected Plant Extracts for Anti Tyrosinase Activity. *J. Nat. Remedies* **2016**, *16*, 18. [CrossRef]

17. Laddha, A.P.; Kulkarni, Y.A. Tannins and vascular complications of Diabetes: An update. *Phytomedicine* **2019**, *56*, 229–245. [CrossRef] [PubMed]

18. Sillero, L.; Prado, R.; Andrés, M.A.; Labidi, J. Characterisation of bark of six species from mixed Atlantic forest. *Ind. Crops Prod.* **2019**, *137*, 276–284. [CrossRef]

19. Chemat, F.; Abert-Vian, M.; Fabiano-Tixier, A.S.; Strube, J.; Uhlenbrock, L.; Gunjevic, V.; Cravotto, G. Green extraction of natural products. Origins, current status, and future challenges. *Trends Analyt. Chem.* **2019**, *118*, 248–263. [CrossRef]

20. Razzaghi, S.E.; Arabhosseini, A.; Turk, M.; Soubrat, T.; Cendres, A.; Kianmehr, M.H.; Perino, S.; Chemat, F. Operational efficiencies of six microwave based extraction methods for orange peel oil. *J. Food Eng.* **2019**, *241*, 26–32. [CrossRef]

21. Reyes-Ocampo, I.; Córdova-Aguilar, M.S.; Guzmán, G.; Blancas-Cabrera, A.; Ascanio, G. Solvent-free mechanical extraction of Opuntia ficus-indica mucilage. *J. Food Process. Eng.* **2019**, *42*, 12954. [CrossRef]

22. Cendres, A.; Hoerlé, M.; Chemat, F.; Renard, C.M. Different compounds are extracted with different time courses from fruits during microwave hydrodiffusion: Examples and possible causes. *Food Chem.* **2014**, *154*, 179–186. [CrossRef] [PubMed]

23. Rodríguez-Seoane, P.; Díaz-Reinoso, B.; Muñoz, G.-; Portela, C.F.D.A.; Domínguez, H. Innovative technologies for the extraction of saccharidic and phenolic fractions from *Pleurotus eryngii*. *LWT* **2019**, *101*, 774–782. [CrossRef]

24. Benmoussa, H.; Elfalleh, W.; He, S.; Romdhane, M.; Benhamou, A.; Chawech, R. Microwave hydrodiffusion and gravity for rapid extraction of essential oil from Tunisian cumin (*Cuminum cyminum* L.) seeds: Optimization by response surface methodology. *Ind. Crop. Prod.* **2018**, *124*, 633–642. [CrossRef]

25. Darvishi, H.; Azadbakht, M.; Rezaeiasl, A.; Farhang, A. Drying characteristics of sardine fish dried with microwave heating. *J. Saudi Soc. Agric. Sci.* **2013**, *12*, 121–127. [CrossRef]

26. Singleton, V.L.; Rossi, J.A. Colorimety of total phenolics with phosphomolybdic-phosphotungstic acid reagents. *Am. J. Enol Viticult.* **1965**, *16*, 144–148.

27. Re, R.; Pellegrini, N.; Proteggente, A.; Pannala, A.; Yang, M.; Rice-Evans, C. Antioxidant activity applying an improved ABTS radical cation decolorization assay. *Free. Radic. Boil. Med.* **1999**, *26*, 1231–1237. [CrossRef]

28. Benzie, I.F.; Strain, J. The Ferric Reducing Ability of Plasma (FRAP) as a Measure of "Antioxidant Power": The FRAP Assay. *Anal. Biochem.* **1996**, *239*, 70–76. [CrossRef]

29. Gadow, A.; Joubert, E.; Hansmann, C. Comparison of the antioxidant activity of rooibos tea (*Aspalathus linearis*) with green, oolong and black tea. *Food Chem.* **1997**, *60*, 73–77. [CrossRef]

30. Khosa, M.K.; Chatha, S.A.S.; Hussain, A.I.; Zia, K.M.; Riaz, H.; Aslam, K. Spectrophotometric quantification of antioxidant phytochemicals in juices from four different varieties of Citrus limon indigenous to Pakistan. *J. Chem. Soc. Pak.* **2011**, *33*, 188–192.

31. Liyanaarachchi, G.D.; Samarasekera, J.K.R.R.; Mahanama, K.R.R.; Hemalal, K.D.P. Tyrosinase, elastase, hyaluronidase, inhibitory and antioxidant activity of Sri Lankan medicinal plants for novel cosmeceuticals. *Ind. Crops Prod.* **2018**, *111*, 597–605. [CrossRef]

32. Chiari, M.; Joray, M.; Ruiz, G.; Palacios, S.; Carpinella, M. Tyrosinase inhibitory activity of native plants from central Argentina: Isolation of an active principle from *Lithrea molleoides*. *Food Chem.* **2010**, *120*, 10–14. [CrossRef]

33. Balboa, E.M.; Rivas, S.; Moure, A.; Domínguez, H.; Parajó, J.C. Simultaneous Extraction and Depolymerization of Fucoidan from *Sargassum muticum* in Aqueous Media. *Mar. Drugs* **2013**, *11*, 4612–4627. [CrossRef] [PubMed]

34. Garrote, G.; Domínguez, H.; Parajó, J.C. Manufacture of xylose-based fermentation media from corncobs by posthydrolysis of autohydrolysis liquors. *Appl. Biochem. Biotechnol.* **2001**, *95*, 195–208. [CrossRef]

35. Bradford, M.M. A rapid and sensitive method for the quantitation of microgram quantities of protein utilizing the principle of protein-dye binding. *Anal. Biochem.* **1976**, *72*, 248–254. [CrossRef]

36. Kamal, I. Identification and Extraction Kinetics of Lipids Using UV Spectrophotometry. Technical Report. Available online: https://www.researchgate.net/publication/319128233_Identification_and_Extraction_Kinetics_of_Lipids_Using_UV_Spectrophotometry (accessed on 28 August 2017).

37. Kaur, C.D.; Saraf, S. In vitro sun protection factor determination of herbal oils used in cosmetics. *Pharmacognosy Res.* **2010**, *2*, 22–25.

38. Mansur, J.S.; Breder, M.N.R.; Mansur, M.C.A.; Azulay, R.D. Determinação do fator de proteção solar por espectrofotometría. *An. Bras. Dermatol.* **1986**, *61*, 121–124.

39. Sayre, R.M.; Agin, P.P.; Levee, G.J.; Marlowe, E. A Comparison Of In Vivo And In Vitro Testing of Sunscreening Formulas. *Photochem. Photobiol.* **1979**, *29*, 559–566. [CrossRef]

40. Balboa, E.M.; Soto, M.L.; Nogueira, D.R.; González-López, N.; Conde, E.; Moure, A.; Vinardell, M.P.; Mitjans, M.; Domínguez, H. Potential of antioxidant extracts produced by aqueous processing of renewable resources for the formulation of cosmetics. *Ind. Crops Prod.* **2014**, *58*, 104–110. [CrossRef]

41. Dutra, E.A.; Oliveira, D.A.G.D.C.; Kedor-Hackmann, E.R.M.; Santoro, M.I.R.M. Determination of sun protection factor (SPF) of sunscreens by ultraviolet spectrophotometry. *Revista Brasileira de Ciências Farmacêuticas* **2004**, *40*, 381–385. [CrossRef]

42. Scheffler, S.L.; Wang, X.; Huang, L.; Gonzalez, F.S.-M.; Yao, Y. Phytoglycogen Octenyl Succinate, an Amphiphilic Carbohydrate Nanoparticle, and ε-Polylysine To Improve Lipid Oxidative Stability of Emulsions. *J. Agric. Food Chem.* **2010**, *58*, 660–667. [CrossRef] [PubMed]

43. Chemat, F.; Vian, M.; Franco, V. Microwave Hydrodiffusion for Isolation of Natural Products. European Patent EP1,955,749A1, 13 August 2008.

44. Makanjuola, S.A. Influence of particle size and extraction solvent on antioxidant properties of extracts of tea, ginger, and tea-ginger blend. *Food Sci. Nutr.* **2017**, *5*, 1179–1185. [CrossRef]

45. López-Hortas, L.; Gely, M.; Falqué, E.; Domínguez, H.; Torres, M.D. Alternative environmental friendly process for dehydration of edible *Undaria pinnatifida* brown seaweed by microwave hydrodiffusion and gravity. *J. Food Eng.* **2019**, *261*, 15–25. [CrossRef]

46. Ekezie, F.-G.C.; Sun, D.-W.; Cheng, J.-H. Acceleration of microwave-assisted extraction processes of food components by integrating technologies and applying emerging solvents: A review of latest developments. *Trends Food Sci. Technol.* **2017**, *67*, 160–172. [CrossRef]

47. Iqbal, S.; Younas, U.; Chan, K.W.; Saeed, Z.; Shaheen, M.A.; Akhtar, N.; Majeed, A. Growth and antioxidant response of *Brassica rapa var. rapa* L. (turnip) irrigated with different compositions of paper and board mill (PBM) effluent. *Chemosphere* **2013**, *91*, 1196–1202. [CrossRef] [PubMed]

48. Rakic, S.; Petrović, S.; Kukić, J.; Jadranin, M.; Tesevic, V.; Povrenovic, D.; Siler-Marinkovic, S. Influence of thermal treatment on phenolic compounds and antioxidant properties of oak acorns from Serbia. *Food Chem.* **2007**, *104*, 830–834. [CrossRef]

49. Cipolatti, E.P.; Remedi, R.D.; Sá, C.D.S.; Rodrigues, A.B.; Ramos, J.M.G.; Burkert, C.A.V.; Furlong, E.B.; Burkert, J.F.D.M. Use of agroindustrial byproducts as substrate for production of carotenoids with antioxidant potential by wild yeasts. *Biocatal. Agric. Biotechnol.* **2019**, *20*, 101208. [CrossRef]

50. Neha, P.; Pandey-Rai, S. Biochemical activity and therapeutic role of antioxidants in plants and humans. In *Plants as a Source of Natural Antioxidants*; CABI Publishing: Wallingford, UK, 2014; pp. 191–224.

51. Frede, K.; Schreiner, M.; Baldermann, S. Light quality-induced changes of carotenoid composition in pak choi Brassica rapa ssp. chinensis. *J. Photochem. Photobiol. B Boil.* **2019**, *193*, 18–30. [CrossRef] [PubMed]

52. Bessa, L.C.; Ferreira, M.C.; Rodrigues, C.E.; Batista, E.A.; Meirelles, A.J. Simulation and process design of continuous countercurrent ethanolic extraction of rice bran oil. *J. Food Eng.* **2017**, *202*, 99–113. [CrossRef]

53. Page, J.C.; Arruda, N.P.; Freitas, S.P. Crude ethanolic extract from spent coffee grounds: Volatile and functional properties. *Waste Manag.* **2017**, *69*, 463–469. [CrossRef]

54. Thavarajah, D.; Thavarajah, P.; Abare, A.; Basnagala, S.; Lacher, C.; Smith, P.; Combs, G.F. Mineral micronutrient and prebiotic carbohydrate profiles of USA-grown kale (*Brassica oleracea* L. var. acephala). *J. Food Compos. Anal.* **2016**, *52*, 9–15. [CrossRef]

55. Rachmawati, H.; Sundari, S.; Nabila, N.; Tandrasasmita, O.M.; Amalia, R.; Siahaan, T.J.; Tjandrawinata, R.R.; Ismaya, W.T. Orf239342 from the mushroom *Agaricus bisporus* is a mannose binding protein. *Biochem. Biophys. Res. Commun.* **2019**, *515*, 99–103. [CrossRef] [PubMed]

56. Zhang, M.; Liu, Y.; Song, C.; Ning, J.; Cui, Z. Characterization and functional analysis of a novel mannose-binding lectin from the swimming crab *Portunus trituberculatus*. *Fish Shellfish. Immunol.* **2019**, *89*, 448–457. [CrossRef] [PubMed]

57. Adekunte, A.; Tiwari, B.; Cullen, P.; Scannell, A.; O'Donnell, C. Effect of sonication on colour, ascorbic acid and yeast inactivation in tomato juice. *Food Chem.* **2010**, *122*, 500–507. [CrossRef]

58. Afonso, C.; Hirano, R.; Gaspar, A.; Chagas, E.; Carvalho, R.; Silva, F.; Leonardi, G.; Lopes, P.; Silva, C.; Yoshida, C. Biodegradable antioxidant chitosan films useful as an anti-aging skin mask. *Int. J. Boil. Macromol.* **2019**, *132*, 1262–1273. [CrossRef] [PubMed]

59. Conesa, A.; Manera, F.; Brotons, J.; Fernandez-Zapata, J.; Simón, I.; Simón-Grao, S.; Alfosea-Simón, M.; Nicolás, J.M.; Valverde, J.; García-Sanchez, F. Changes in the content of chlorophylls and carotenoids in the rind of Fino 49 lemons during maturation and their relationship with parameters from the CIELAB color space. *Sci. Hortic.* **2019**, *243*, 252–260. [CrossRef]

60. Bom, S.; Jorge, J.; Ribeiro, H.; Marto, J. A step forward on sustainability in the cosmetics industry: A review. *J. Clean. Prod.* **2019**, *225*, 270–290. [CrossRef]

61. Chiocchio, I.; Mandrone, M.; Sanna, C.; Maxia, A.; Tacchini, M.; Poli, F. Screening of a hundred plant extracts as tyrosinase and elastase inhibitors, two enzymatic targets of cosmetic interest. *Ind. Crops Prod.* **2018**, *122*, 498–505. [CrossRef]

62. Delgado-Outeiriño, I.; Araujo-Nespereira, P.; Cid-Fernández, J.; Mejuto, J.C.; Martínez-Carballo, E.; Simal-Gandara, J. Behaviour of thermal waters through granite rocks based on residence time and inorganic pattern. *J. Hydrol.* **2009**, *373*, 329–336. [CrossRef]

63. Nurjanah; Luthfiyana, N.; Hidayat, T.; Nurilmala, M.; Anwar, E. Utilization of seaweed porridge *Sargassum* sp. and *Eucheuma cottonii* as cosmetic in protecting skin. *IOP Conf. Ser. Earth Environ. Sci.* **2019**, *278*, 012055.

64. Ospina, M.; Montaña-Oviedo, K.; Díaz-Duque, Á; Toloza-Daza, H.; Narváez-Cuenca, C.-E. Utilization of fruit pomace, overripe fruit, and bush pruning residues from Andes berry (*Rubus glaucus* Benth) as antioxidants in an oil in water emulsion. *Food Chem.* **2019**, *281*, 114–123. [CrossRef] [PubMed]

65. Pandolsook, S.; Kupongsak, S. Storage stability of bleached rice bran wax organogels and water-in oil emulsions. *J. Food Meas. Charact.* **2019**, *13*, 431–443. [CrossRef]

66. Houlden, R.J. Viscosity vs. rheology: Why it is important to formulators. Personal Care 2017. Available online: https://www.personalcaremagazine.com/story/24670/viscosity-vs-rheology-why-it-is-important-to-formulators (accessed on 22 November 2017).

67. Houlden, R.J. *The Influence of Rheology on Sunscreen Performance and SPF—Are Highly Thixtotropic Products Not Providing Enough Protection?* TKS Publisher: Milan, Italy, 2018; Volume 13.

68. López-Hortas, L.; Conde, E.; Falqué, E.; Domínguez, H.; Torres, M.D. Preparation of Hydrogels Composed of Bioactive Compounds from Aqueous Phase of Artichoke Obtained by MHG Technique. *Food Bioprocess Technol.* **2019**, *12*, 1304–1315. [CrossRef]

69. López-Hortas, L.; Conde, E.; Falqué, E.; Domínguez, H.; Torres, M. Recovery of aqueous phase of broccoli obtained by MHG technique for development of hydrogels with antioxidant properties. *LWT* **2019**, *107*, 98–106. [CrossRef]

70. Nooeaid, P.; Chuysinuan, P.; Techasakul, S. Alginate/gelatine hydrogels: characterisation and application of antioxidant release. *Green Mater.* **2017**, *5*, 1–12. [CrossRef]

71. Pierson, J.T.; Perino-Issartier, S.; Ruiz, K.; Cravotto, G.; Chemat, F. Laboratory to pilot-scale optimization of microwave hydrodiffusion and gravity: Solvent-free polyphenols extraction from lettuce. *Food Chem.* **2016**, *204*, 108–114.

 *molecules*

Article

Immunomodulatory Effects of the *Meretrix Meretrix* Oligopeptide (QLNWD) on Immune-Deficient Mice

Wen Zhang [1,†], Lei Ye [1,†], Fenglei Wang [2], Jiawen Zheng [1], Xiaoxiao Tian [1], Yan Chen [1,*] , Guofang Ding [1] and Zuisu Yang [1]

[1] Zhejiang Provincial Engineering Technology Research Center of Marine Biomedical Products, School of Food and Pharmacy, Zhejiang Ocean University, Zhoushan 316022, China; zhangwen1225z@163.com (W.Z.); Gamma59777@163.com (L.Y.); jwzheng1996@163.com (J.Z.); TIANXIAOXIAO0208@163.com (X.T.); dinggf2007@163.com (G.D.); abc1967@126.com (Z.Y.)
[2] Zhejiang Hailisheng Group Co., Ltd., Zhoushan 316021, China; 662199.w@163.com
*   Correspondence: cyancy@zjou.edu.cn; Tel.: +86-0580-226-0600; Fax: +86-0580-254-781
†   These authors contributed equally to this work.

Received: 4 November 2019; Accepted: 3 December 2019; Published: 5 December 2019

Abstract: The aim of this study was to explore the immunomodulatory effects of the *Meretrix meretrix* oligopeptide (MMO, QLNWD) in cyclophosphamide (CTX)-induced immune-deficient mice. Compared to untreated, CTX-induced immune-deficient mice, the spleen and thymus indexes of mice given moderate (100 mg/kg) and high (200 mg/kg) doses of MMO were significantly higher ($p < 0.05$), and body weight loss was alleviated. Hematoxylin-eosin (H&E) staining revealed that MMO reduced spleen injury, thymus injury, and liver injury induced by CTX in mice. Furthermore, MMO boosted the production of immunoglobulin G (IgG) and hemolysin in the serum and promoted the proliferation and differentiation of spleen T-lymphocytes. Taken together, our findings suggest that MMO plays a vital role in protection against immunosuppression in CTX-induced immune-deficient mice and could be a potential immunomodulatory candidate for use in functional foods or immunologic adjuvants.

Keywords: *Meretrix meretrix* oligopeptides; cyclophosphamide; immunomodulatory; immune-deficient mice

1. Introduction

Immunoregulation can be broadly divided into positive regulation and negative regulation, both of which are the result of complex regulation of the immune system. Sometimes regulation in only one direction is triggered, but most immune regulation is bidirectional in order to maintain a stable steady-state. Immunomodulators can be classified into three general types: immunopotentiators, immunosuppressants, and two-way immunomodulators [1–3]. When the body experiences diseases or immune abnormalities, the application of immunomodulators can restore immune function to normal. There are many types of immunomodulators, such as bacterial preparations (e.g., lipopolysaccharide (LPS)), chemical preparations (e.g., cyclophosphamide (CTX)), and biochemical preparations (e.g., thymosin) [4]. However, some chemical immunomodulators have serious side effects, which not only have a specific inhibitory effect on the cause of immune diseases, but also have general inhibitory effects on normal tissue cells [5]. Inflammation, infection, tumors, organ bleeding, and loss of pregnancy have all been reported as being induced after the administration of chemical immunomodulators [6]. Untreated chronic inflammation, however, inhibits natural killer (NK) cells and T cells, which are key participants in the immune system, and limits the success of immunotherapy [7]. More recently, immunomodulators from natural extracts have attracted much attention in the field due to their lesser side-effects when used in humans [8,9]. For example, Hong et al. [10] showed that *Cervus nippon mantchuricus* extract (NGE) has immuno-enhancing effects on RAW264.7 macrophage cells in immunosuppressed mice. Purified leaf extracts of *Melia azedarach* L. (CDM) exerted anti-herpetic activity, inhibited NFκB translocation to the nucleus, and modulated both interleukin (IL)-6 and tumor necrosis factor-alpha (TNF-α) responses in macrophages in one recent study [11]. Therefore, further exploration of natural and effective immunomodulators with lesser side effects seems to be a very worthwhile research pursuit.

Bioactive peptides are small proteins, composed of amino acids, which often have unique physiological functions not possessed by large proteins or their constituent amino acids, such as antibacterial, antiviral, anti-oxidant, antifungal, calcium-binding, or anti-tumor properties [12–14]. Moreover, many bioactive peptides can be absorbed and digested even more quickly than free amino acids, and thus have become popular research topics and promising functional factors in the international food industry [15]. As the most common kind of bioactive peptide, immunologically active peptides stimulate the proliferation of lymphocytes, enhance the phagocytic abilities of macrophages, improve the body's resistance to external pathogens, and generally enhance the body's immunity to infection. Recently, such immunoregulatory peptides have attracted much research attention. For example, Yang et al. [16] reported that a marine oligopeptide from chum salmon could significantly enhance the capacity of lymphocyte proliferation in mice. Gao et al. [17] reported that collagen hydrolysates from yak bones exhibited immunomodulatory effects on CTX-induced immunosuppressed mice by increasing both innate and adaptive immunity. Li et al. [18] reported that a novel pentapeptide (RVAPEEHPVEGRYLV) from *Cyclina sinensis* could stimulate macrophage activity to activate the NFκB signaling pathway, and further in vivo studies revealed that this novel pentapeptide has immunomodulatory effects on CTX-induced immunosuppression in mice [19].

In a previous study of ours, an oligopeptide (QLNWD) was purified from the hydrolysate of *Meretrix meretrix* oligopeptide (MMO), and was shown to have the ability to aid in reversing the effects of nonalcoholic fatty liver disease (NAFLD) in mice [20]. We also investigated the immunomodulatory effect of this oligopeptide in vitro [21], and our results indicated that MMO has the effect of promoting the activation of RAW264.7 cells and the potential to enhance the non-specific immunity. However, the immunoregulatory activity of this oligopeptide in vivo is unknown. The aim of the present study was to explore the immunomodulatory effects of MMO on mice with CTX-induced immunosuppression in vivo. The effects of MMO on the thymus and spleen indexes of the mice were investigated, as well as morphological changes to their spleens, thymuses, and livers as observed microscopically using hematoxylin-eosin (H&E) staining. The stimulation index change in spleen T-lymphocytes was also determined in the present study. This study will provide a foundation for the further development of MMO as an immunopotentiator.

2. Results and Discussion

2.1. Comparison of Body Weight

The body weight of mice is a direct indicator of their physical condition. Previous evidence has shown that weight recovery can effectively increase the number of T-cell subsets and macrophages, which are vital components of the murine immune system [22–24]. As shown in Figure 1, it was observed that the weights of the mice in the positive drug or MMO-treated groups were significantly reduced when compared to the control group in the five days prior to the commencement of the study. Over the next 10 days, the mice in both the positive control group and the MMO-treated group saw marked increases in their body weights. However, the body weight of mice in the disease model became stable after 7 days, as it seems that CTX can cause immunodeficiency in mice, resulting in reduced appetite. This result indicated that MMO had an effect on alleviating the degree of immunosuppression induced by CTX on the mice.

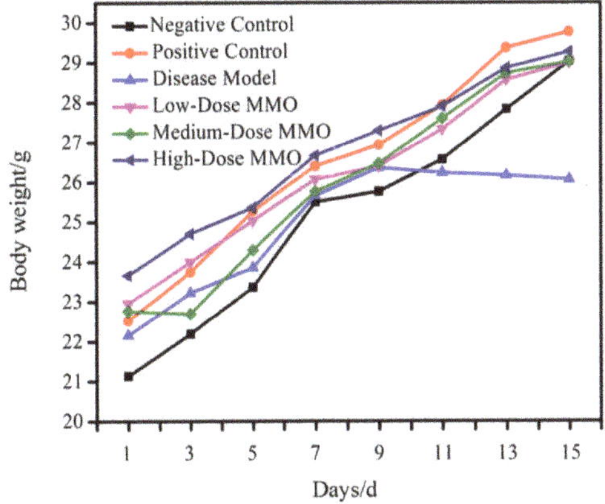

Figure 1. The body weight changes of immunosuppressed study mice. Negative control: saline; positive control: 25 mg/kg of levamisole; disease model: 80 mg/kg of CTX; low-dose MMO: 50 mg/kg of MMO; medium-dose MMO: 100 mg/kg of MMO; and high-dose MMO: 200 mg/kg of MMO.

2.2. Thymus and Spleen Indexes

The thymus and spleen are representative immune organs. The thymus is one of the primary lymphoid organs [25], and the innate and adaptive immune responses to antigens and pathogens are initiated by the spleen, which is considered to be an important organ for assessing immune system function [26]. The thymus and spleen indexes can thus be used to roughly estimate the strength of immune function, which is a superficial and lagging indicator [27]. Our results showed that the spleen and thymus indexes of the model group were visibly reduced ($p < 0.05$) when compared to the negative control group. The spleen indexes of the middle (100 mg/kg) and high (200 mg/kg) MMO groups were significantly higher than those of the model group ($p < 0.05$), which was similar for their thymus indexes as well (Figure 2). These results indicate that MMO may be able to effectively alleviate the atrophy of both the spleen and the thymus caused by CTX.

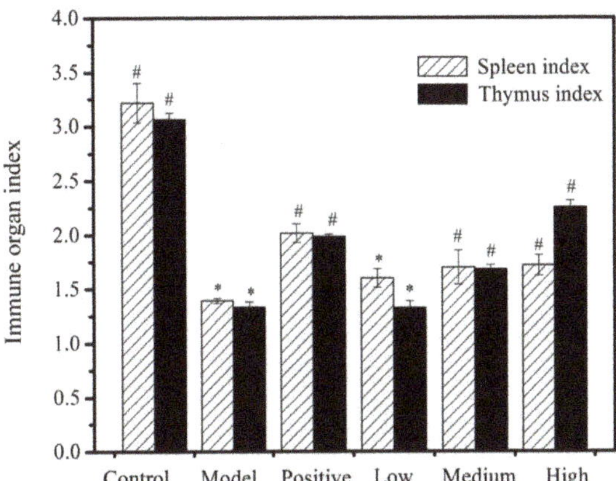

Figure 2. MMO-induced changes to the immune organ indexes of mice. *, a significant difference when compared to the negative control ($p < 0.05$); #, a significant difference when compared to the disease model, $p < 0.05$ values were considered to be statistically significant.

2.3. Morphological Observations of Mouse Organs

To gain further insight into the effectiveness of MMO on CTX-induced immunosuppression in the mice, H&E staining was used to observe subtle morphological changes to the spleens, thymus glands, and livers of the mice in each group (Figures 3–5).

The spleen is the largest immune organ of the body, accounting for 25% of the total lymphoid tissue, and also contains a large number of lymphocytes, dendritic cells, neutrophils, natural killer cells, and macrophages [28]. As shown in Figure 3, there was a clear dividing line between the red and white medulla in the normal group (Figure 3A). The splenic corpuscle in the white medulla was nearly round, the lymphatic sheath structure around the artery was complete, the splenic cord in the red medulla was connected, and the splenic sinus was obvious. In the model group, the boundary between the white pulp and the red pulp was blurred and the splenic corpuscle in the white pulp was scattered (Figure 3B). Compared to the normal group, the lymphatic sheath around the artery was thinner and the area of the red pulp was smaller, which suggested that CTX may have damaged the T and B cells in the spleen, significantly reduced the lymphoid tissue, and led to overall atrophy of the spleen. In the positive control group, multiple intact splenic corpuscles were seen, with an enlarged white medullary margin and thickened lymphatic sheath around the central artery (Figure 3C). The general structure of the splenic corpuscle was observed in the low-dose MMO group, but the boundary between the red and white medulla was still not obvious (Figure 3D). In the medium-dose MMO group, the marginal area of white pulp could be observed (Figure 3E). Red and white medulla were clearly observed in the high-dose MMO group and the white medulla margins were widened (Figure 3F). Overall, MMO gradually returned the spleen structure to an organizational form similar to that of the normal group. The white pulp part, for example, became clearer and, for the high-dose MMO group in particular, presented with a shape very similar to that of the positive control group. These results suggest that MMO can restore lymphocyte white marrow, increase T and B cells in marginal regions, and reduce CTX-induced spleen cell apoptosis in mice [29].

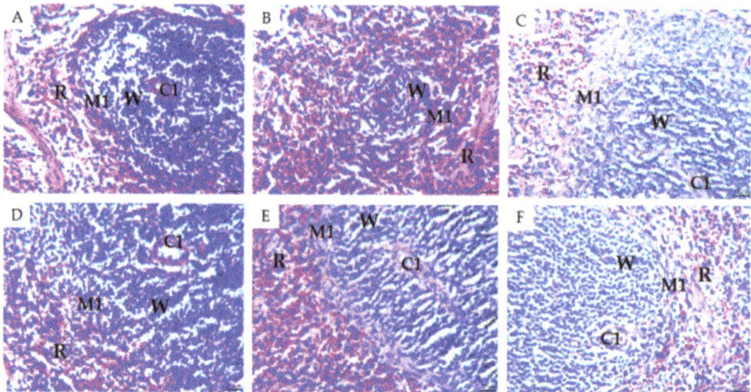

Figure 3. H&E staining of mouse spleen (×400). R: red medulla; W: white medulla; M1: marginal area; C1: central artery. (**A**) Negative control; (**B**) disease model; (**C**) positive control; (**D**) low-dose (50 mg/kg) MMO; (**E**) medium-dose (100 mg/kg) MMO; and (**F**) high-dose (200 mg/kg) MMO.

Thymus atrophy also showed a similar trend in all of the mice. The cortex of the thymus contains thymocytes, which produce thymosin that can stimulate the proliferation and differentiation of T-lymphocytes, activate the major histocompatibility complex (MHC) colony factor transmitting signal [30], and accelerate the presentation of antigens. In the normal group, cortical and medullary structures were clear and distinct and obvious thymus bodies were observed in the medulla (Figure 4A). In the model group, the cortex and medulla were intercalated, the thymus corpuscle was shrunken and unclear in the visual field, the cortical area was smaller, and the number of T-lymphocytes was significantly reduced (Figure 4B), which all indicate significant immunosuppression when compared to the normal group. Cells and T-lymphocytes in the thymus of the positive control group were significantly increased when compared to the negative control group and multiple thymus corpuscles were observed in the visual field (Figure 4C). The cortex and medulla of the low-dose MMO group still could not be distinguished, but there was an increase in T cells in the cortex (Figure 4D). In the medium-dose MMO group, the cortex and medulla could be distinguished only roughly (Figure 4E). In the high-dose MMO group, however, the cortex and medulla were distinct and T-lymphocytes were significantly increased (Figure 4F)—a similar morphology to that of the positive control group. With increasing doses of MMO, cortical thymocytes increased, which demonstrated that MMO could activate the immune response and reduce the thymus injury induced by CTX [30].

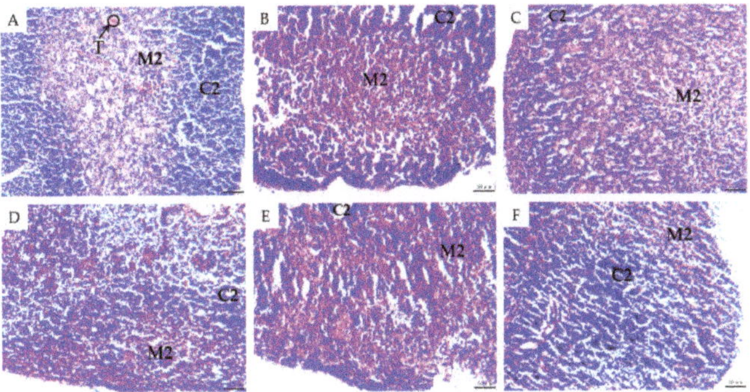

Figure 4. H&E staining of mice thymus (×200). C2: cortical; M2: medulla; T: thymus corpuscle. (**A**) Negative control; (**B**) disease model; (**C**) positive control; (**D**) low-dose (50 mg/kg) MMO; (**E**) medium-dose (100 mg/kg) MMO; and (**F**) high-dose (200 mg/kg) MMO.

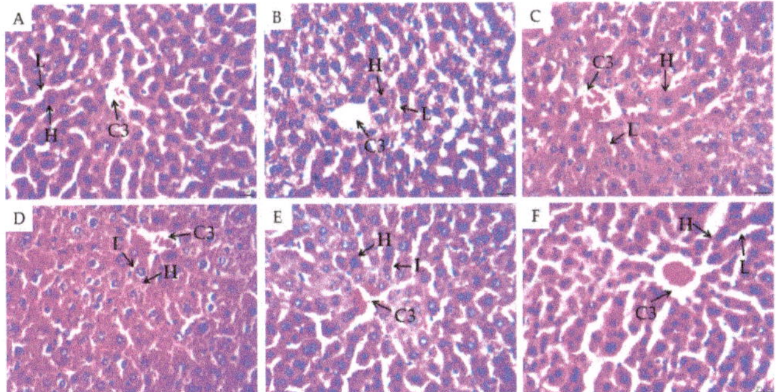

Figure 5. H&E staining of mice livers (×400). C3: central vein; H: hepatic cord; L: liver sinusoidal. (**A**) Negative control; (**B**) disease model group; (**C**) positive control; (**D**) low-dose (50 mg/kg) MMO; (**E**) medium-dose (100 mg/kg) MMO; and (**F**) high-dose (200 mg/kg) MMO.

The liver is the central hub of the body's metabolism, with functions such as detoxification and hematopoiesis [31]. To investigate whether there was an effect on the liver after using immunosuppressive agents and MMO, the histological structure of the mouse liver was observed. The hepatic lobule structure of the negative control group was clear and complete, with radial hepatic cords that radiated out in all directions and were arranged neatly around the central vein. Morphology of the hepatocytes was also regular and liver sinusoidal structures were observed (Figure 5A). In the disease model group, the hepatic cord was ruptured and disordered and degeneration and necrosis of hepatocytes was evidenced by a reduction in vacuoles and even absence of part of the nucleus. The structure of the liver sinusoids was not obvious, indicating that the cytotoxicity of CTX caused damage to the mouse liver (Figure 5B). In the positive control group, the hepatic cord around the central vein recovered to the radial structure and the hepatocyte nucleus also showed a round shape (Figure 5C). The liver morphologies of the medium-dose and the high-dose MMO groups both bore a close resemblance to that of the positive control group (Figure 5E,F). Our results showed that MMO can effectively reduce the cytotoxicity of CTX-induced liver injury in mice.

2.4. Serum Immunoglobulin G (IgG) Levels

IgG is one of the most abundant proteins in human serum, accounting for about 10–20% of plasma protein [32]. Detection of IgG levels can help to indirectly judge the immune function of the body [33]. Vikas et al. [34] explored the changes in IgG levels in mice treated with galactose. The results showed that the IgG concentration in the galactose-treated mice was higher than that of the normal group, indicating that galactose had the potential to upregulate IgG production. The effect of MMO on IgG content in mice serum is shown in Figure 6. The IgG level in the model group was markedly reduced when compared to the negative control group ($p < 0.05$). From the doses of 50 mg/kg to 200 mg/kg, the MMO-treated groups seemed to have significant dose-dependent increases in IgG levels ($p < 0.05$), when compared to the disease model group. Moreover, the IgG levels in the high-dose MMO group were higher than the negative control and close to the positive control group, which suggests that high doses of MMO have a very beneficial effect on the restoration of serum immunoglobulins in immunocompromised mice.

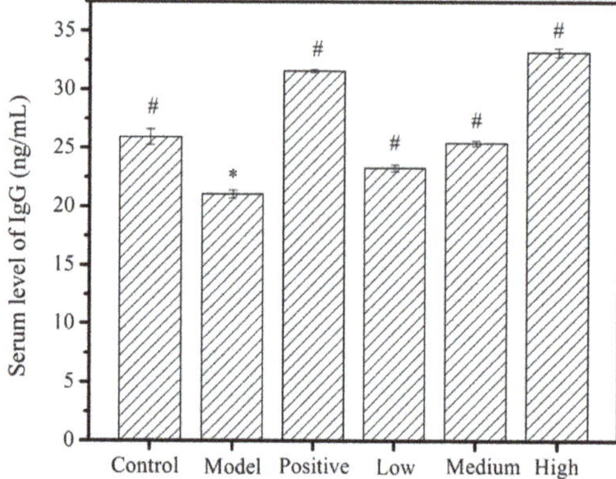

Figure 6. The effects of MMO on immunoglobulin G (IgG) content in mouse serum. *, a significant difference when compared to the negative control, $p < 0.05$. #, a significant difference when compared to the disease model, $p < 0.05$. $p < 0.05$ values were considered to be statistically significant.

2.5. Serum Hemolysin

Hemolysin reflects the proliferation and differentiation of hemolytic B cells and is one of the main nonspecific indexes used to measure the immune function of the body [35,36]. The half hemolysis value (HC_{50}) and the hemolysin proliferation rate are routinely used to evaluate the effects of natural extract products on humoral immunity in mice [36]. As shown in Table 1, in contrast to the negative control group, the HC_{50} of the model group dropped by 0.68 ± 0.05. However, the HC_{50} levels of the MMO-treated groups (50, 100, 200 mg/kg) were raised by 0.26 ± 0.05, 1.69 ± 0.02, and 3.20 ± 0.05, respectively. The proliferation rate of hemolysin in the disease model group was −0.78% ± 0.05 when compared to the negative control group, which indicated that serum hemolysin was inhibited by CTX. However, the proliferation rates of the MMO-treated groups exceeded that of the negative control group across the board, suggesting a supra-accelerating effect of MMO on CTX-damaged mice. Similarly, Pan et al. [35] reported that milk protein hydrolysate (MPH) increased immunological function by triggering hemolysin formation in mice. Liu et al. [37] showed that cottonseed meal oligopeptide (PFC) significantly increased the HC_{50} levels in mice by 1.39 ± 0.45, 2.59 ± 0.20, and 2.46 ± 0.41 when given doses of 5 mg/mL, 10 mg/mL, and 20 mg/mL, respectively. Our results were consistent with these findings and indicated that MMO has the effect of alleviating immunosuppression induced by CTX in mice.

Table 1. The effect of MMO on HC_{50} and the hemolysin proliferation rate.

Group	HC_{50}	Proliferation Rate %
Negative Control	87.58 ± 0.05	0
Diseased Model	86.90 ± 0.05	−0.78 ± 0.05
Positive Control	91.39 ± 0.03 *	4.35 ± 0.03 *
Low-dose	87.84 ± 0.10	0.29 ± 0.10
Medium-dose	89.27 ± 0.07	1.92 ± 0.07
High-dose	90.78 ± 0.05 *	3.65 ± 0.05 *

Note: * indicates a significant difference over the Negative Control group.

2.6. T Lymphocyte Assessment

As the main cells in both the thymus and the spleen, T-lymphocytes can assist B cells to produce antibodies, kill target cells, and promote mitogen responses [38,39]. Relevant studies have reported that mitogens, such as Concanavalin A (ConA) and phytohemagglutinin (PHA), can stimulate lymphocytes to release a wide variety of cytokines in vitro as well as induce the simultaneous stimulatory and inhibitory activities of different T cell populations [40]. Evidence has shown that one effect of ConA stimulation of T-lymphocytes may be to enhance endocytosis of the cell membrane and studies have speculated that cell density and cell contact area are associated with the stimulation of ConA, reaching a peak of growth at 24 h [41,42]. In this study, we used ConA to stimulate spleen T-lymphocytes extracted from the spleens of each group of mice and observed the stimulation effects after 24 h. We observed that the stimulation of T-lymphocytes in the model group was lower than in the negative control, which indicated that CTX inhibited the responses of T cells in the lymphatic systems of those mice. The net proliferation of T-lymphocytes in the low-dose (50 mg/kg) MMO group was similar to that of the model group. However, the net proliferation of T-lymphocytes in the high-dose MMO-treated group was higher than that of the negative control group, indicating that MMO stimulated the proliferation of T-lymphocytes and reduced the inhibitory effects of CTX on T-lymphocytes in those mice (Figure 7). Combined with the H&E staining results, we speculated that spleen atrophy was reduced by MMO and that the activity of T-lymphocytes was increased in CTX-immunocompromised mice.

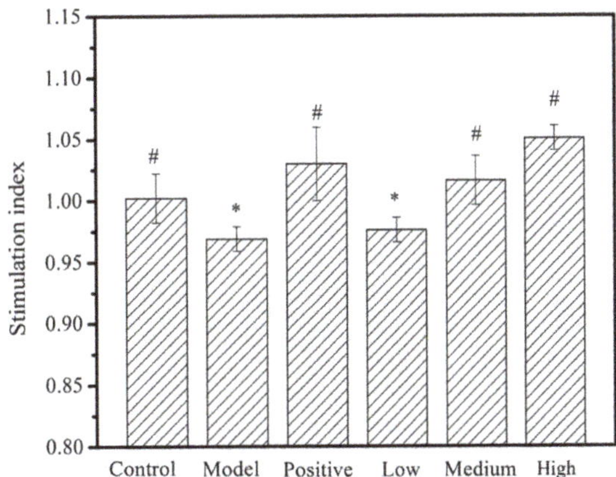

Figure 7. The stimulation index change induced by MMO in spleen T-lymphocytes from mice. *, a significant difference when compared to the negative control ($p < 0.05$); #, a significant difference when compared to the disease model ($p < 0.05$). $p < 0.05$ values were considered to be statistically significant.

3. Materials and Methods

3.1. Animals

Sixty male ICR mice (20–23 g) were provided by the Experiment Animal Center of Zhejiang Province (certificate no SCXK 2014-0001). All the mice were kept under conventional and uniform conditions at 22 °C. The study proceeded after the mice were given seven days to acclimatize to their new environment.

3.2. Materials and Chemical Reagents

The MMO (QLNWD) [13] used for the experiments was synthesized by China Peptides Co., Ltd. (Shanghai, China). The CTX and levamisole were provided by Shanghai Yuanye Bio-Technology Co., Ltd. (Shanghai, China). An H&E staining kit was supplied by Nanjing Jiancheng Bioengineering Institute (Jiangsu, China). Sheep red blood cells (SRBC) and guinea pig serum were obtained from Zhengzhou Baiji Biological Engineering Co. Ltd. (Henan, China). A mouse IgG enzyme linked immunosorbent assay (ELISA) kit was purchased from Shanghai Fusheng Industrial Co. Ltd. (Shanghai, China). Hanks' balanced salt solution (HBSS) and ConA were purchased from Solarbio (Beijing, China). NLRP3 rabbit monoclonal antibody was purchased from Cell Signaling Technology (Massachusetts, USA). A 3,3'-diaminobenzidine (DAB) immunohistochemistry color development kit was purchased from BBI Life Science Corporation Co., Ltd. (Shanghai, China). Ammonium-chloride-potassium (ACK) lysis buffer was offered by Beyotime Biotechnology (Shanghai, China).

3.3. Animal Groupings and Treatments

Animal groupings and procedures were performed according to the methods in Zhang et al. [43] with some slight modifications. The mice were randomly divided into 6 groups, each of which contained 10 mice. Each mouse had its body weight recorded, received an intraperitoneal injection of 0.2 mL, and was fed the same weight of feed every day at the same time. The experimental groups were administered three dosage concentrations of MMO: 50 mg/kg body weight (BW) (low dose), 100 mg/kg BW (medium dose), and 200 mg/kg BW (high dose). The negative control group and the disease model group were given normal saline (NS, 0.9% NaCl) injections. The positive control group was protected from the effects of CTX by 2.5 mg/kg BW of levamisole given over 10 days prior to the experiment [44]. On day 1 of the experiment, all groups except the negative control group were injected with 80 mg/kg BW CTX (Table 2, after having fasted without water deprivation for 24 h beforehand [45].

Table 2. Mice groupings and treatments.

Group	Pre-Treatment (10 days)	Treatment (5 days)
	Dose (0.2 mL)	Dose (0.2 mL)
Negative Control	NS	NS
Disease Model	NS	CTX (80 mg/kg BW)
Positive Control	Levamisole (25 mg/kg BW)	CTX (80 mg/kg BW)
Low Dose	MMO (50 mg/kg BW)	CTX (80 mg/kg BW)
Medium Dose	MMO (100 mg/kg BW)	CTX (80 mg/kg BW)
High Dose	MMO (200 mg/kg BW)	CTX (80 mg/kg BW)

3.4. Body Weight and Immune Organ Index Changes

The body weights of all mice were recorded once every other day for 15 days total. Before being sacrificed by cervical dislocation, each mouse was weighed a final time. The immune organs and the spleen and thymus glands were harvested, rinsed using NS, blotted by gauze immediately, and weighed in order to calculate each mouse's immune organ index (IOI) using Equation (1), before finally proceeding to dissection:

$$\text{IOI} = \frac{\text{immune organ weight}}{\text{bodyweight}} \times 100\%. \tag{1}$$

3.5. Histomorphological Observation

Following dissection, the tissues were fixed in 4% paraformaldehyde for 24 h to 48 h, embedded in paraffin, sliced to 5 µm sections, stained using an H&E staining kit, and sealed with neutral gum. The histomorphological changes of the organ tissues of each group were observed under an optical microscope (CX31, Olympus) and photographed with a CCD-NC 6051 photographic system.

3.6. Determination of IgG Serum Content

Blood was sampled from the eyes 24 h after each mouse's last intraperitoneal injection. The serum and plasma were separated using a refrigerated centrifuge (4 °C, 5000 rpm, 5 min). The amount of IgG in the serum was measured by a mouse IgG ELISA kit from Shanghai Fusheng Industrial Co. Ltd.

3.7. Detection of Serum Hemolysin

The mice serum samples were diluted 100-fold in 96-well plates at 100 µL per well. The sample wells were mixed with 5% sheep red blood cells (SRBC) (50 µL) and 10% guinea pig serum (50 µL), while control wells had just 5% SRBC (50 µL) added to them. The 96-well plates were placed in a 37 °C water bath for 30 min, after which the reaction was stopped in ice water and the supernatants were collected and analyzed at 540 nm in a microplate reader (SpectraMax M2, Molecular Devices, San Jose, CA, USA). The half hemolysis value (HC_{50}) and the hemolysin content change showed the change of hemolysin in the serum samples of the mice (Equations (2) and (3)).

$$HC_{50} = \frac{\text{OD value of sample} \times \text{dilution ratio}}{\text{OD value of SRBC}}, \tag{2}$$

$$\text{Proliferation rate} = \frac{(HC_{50} \text{ of sample} - HC_{50} \text{ of control})}{HC_{50} \text{ of control}}. \tag{3}$$

3.8. Proliferation of Spleen T-Lymphocytes

Mice spleen T-lymphocytes were extracted by the method described by Cai et al. [46]. The spleens of the mice were carefully dissected on a sterile bench, washed with HBSS, cleaned of blood and unrelated tissues, and ground on a 200-mesh stainless steel mesh, after which the cells were collected in a clean centrifuge tube. After being centrifuged at 1000 rpm for 5 min, the supernatant was discarded and the pellet was mixed with 2 mL of ammonium-chloride-potassium (ACK) lysis buffer for 5 min, washing three times with HBSS in between each step. After centrifugation under uniform conditions, the remaining cells were resuspended in RPMI-1640 complete medium, cultured in a cell culture incubator for 12 h, and stored for subsequent multiplex reaction experiments.

The T lymphocyte multiplication experiment method described by Ye et al. [47] was adopted for this study as well. The pre-preparation cell suspension was then put into a 96-well plate and the number of cells was 1×10^6 cells/mL. Each concentration was set into four complex wells (200 µL). Then, 10 µL of ConA (5 µg/mL) was added among 2 wells and 10 µL NS added as control. The plate was incubated in an incubator (37 °C, 5% CO_2) for 24 h, to which was added 3-(4,5-dimethyl-2-thiazolyl)-2,5-diphenyl-2-*H*-tetrazolium bromide (MTT) during the 20 h. Finally, it was treated with 150 µL DMSO for 10 min and the absorbance was detected under 490 nm. The T lymphocyte-multiplication extent was represented by the stimulation index (SI), calculated as shown below Equation (4):

$$SI = \frac{\text{OD value of sample well (average)}}{\text{OD value of control (average)}} \tag{4}$$

3.9. Statistical Analysis

The experimental data were analyzed and processed by SPSS 19.0 statistical software. The figures were expressed as mean ± standard deviation (SD), analyzed using a one-way analysis of variance (ANOVA) test, and $p < 0.05$ values were considered to be statistically significant.

4. Conclusions

In general, we have conclusively shown that MMO has immunomodulatory effects on CTX-immunocompromised mice. Compared to the disease model group, 100 mg/kg and 200 mg/kg doses of MMO were shown to significantly increase the spleen and thymus indexes ($p < 0.05$) and alleviate CTX-induced body weight loss in our experimental mice. The spleen immune injuries and thymus injuries induced by CTX were also alleviated in the MMO-treated groups. Furthermore, MMO may increase the levels of IgG and hemolysin in mouse serum and promote the proliferation of spleen T-lymphocytes. Our findings suggest that MMO plays a vital role in protection against immunosuppression in CTX-treated mice. Transcriptomics and proteomics will be used to further reveal its immune regulatory mechanism in our future studies in vitro and in vivo. We hope that our findings will provide a foundation for further study of MMO as an immunoregulatory adjuvant or functional food additive.

Author Contributions: Y.C. conceived and designed the experiments. W.Z., L.Y., F.W., J.Z., X.T., Z.Y., and G.D. performed the statistical analysis of the data. W.Z. and Y.L. wrote the manuscript.

Funding: This work was financially supported by Zhejiang Provincial Natural Science Foundation of China (grant No. LQ18B060004), the National Natural Science Foundation of China (grant No. 21808208) and the Scientific Research Start-up Funds of Zhejiang Ocean University (grant No. 11135090118).

Conflicts of Interest: The authors declare no conflict of interest.

References

1. Deguine, J. New Flavors in Immunomodulation. *Cell* **2018**, *173*, 1553–1555. [CrossRef] [PubMed]
2. Gein, S.V.; Sharav'eva, I.L. Immunomodulating Effects of Cold Stress. *Biol. Bull. Rev.* **2018**, *8*, 482–488. [CrossRef]
3. Yahfoufi, N.; Mallet, J.F.; Graham, E.; Matar, C. Role of probiotics and prebiotics in immunomodulation. *Curr. Opin. Food Sci.* **2018**, *20*, 82–91. [CrossRef]
4. Yao, W.; Wang, F.; Wang, H. Immunomodulation of artemisinin and its derivatives. *Sci. Bull.* **2016**, *61*, 1–8. [CrossRef]
5. Turowski, R.C.; Triozzi, P.L. Application of chemical immunomodulators to the treatment of cancer and AIDS. *Cancer Invest.* **1994**, *12*, 620–643. [CrossRef] [PubMed]
6. Backus, K.M.; Cao, J.; Maddox, S.M. Opportunities and challenges for the development of covalent chemical immunomodulators. *Bioorg. Med. Chem. Lett.* **2019**, *27*, 3421–3439. [CrossRef]
7. Kanterman, J.; Sade-Feldman, M.; Baniyash, M. New insights into chronic inflammation-induced immunosuppression. *Semin. Cancer Biol.* **2012**, *22*, 307–318. [CrossRef]
8. Seyed, M.A. A comprehensive review on Phyllanthus derived natural products as potential chemotherapeutic and immunomodulators for a wide range of human disease. *Biocatal. Agr. Biotechnol.* **2019**, *17*, 529–537. [CrossRef]
9. Tarnawski, M.; Depta, K.; Grejciun, D.; Szelepin, B. HPLC determination of phenolic acids and antioxidant activity in concentrated peat extract—a natural immunomodulator. *J. Pharmaceut. Biomed.* **2006**, *41*, 182–188. [CrossRef]
10. Hong, S.H.; Ku, J.M.; In Kim, H.; Ahn, C.-W.; Park, S.-H.; Seo, H.S.; Shin, Y.C.; Ko, S.-G. The immune-enhancing activity of Cervus nippon mantchuricus extract (NGE) in RAW264.7 macrophage cells and immunosuppressed mice. *Food Res. Int.* **2017**, *99*, 623–629. [CrossRef]
11. Bueno, C.A.; Barquero, A.A.; Di Cónsoli, H.; Maier, M.S.; Alché, L.E. A natural tetranortriterpenoid with immunomodulating properties as a potential anti-HSV agent. *Virus Res.* **2009**, *141*, 47–54. [CrossRef] [PubMed]

12. Song, R.; Wei, R.; Zhang, B.; Yang, Z.; Wang, D. Antioxidant and Antiproliferative Activities of Heated Sterilized Pepsin Hydrolysate Derived from Half-Fin Anchovy (Setipinna taty). *Mar. Drugs* **2011**, *9*, 1142–1156. [CrossRef] [PubMed]

13. Huang, F.; Zhao, S.; Yu, F.; Yang, Z.; Ding, G. Protective Effects and Mechanism of Meretrix meretrix Oligopeptides against Nonalcoholic Fatty Liver Disease. *Mar. Drugs* **2017**, *15*, 31. [CrossRef] [PubMed]

14. Yu, F.; Zhang, Y.; Ye, L.; Tang, Y.; Ding, G.; Zhang, X.; Yang, Z. A novel anti-proliferative pentapeptide (ILYMP) isolated fromCyclina sinensisprotein hydrolysate induces apoptosis of DU-145 prostate cancer cells. *Mol. Med. Rep.* **2018**, *18*, 771–778. [PubMed]

15. Pang, G.C.; Chen, Q.S.; Hu, Z.-H.; Xie, J.B. Bioactive Peptides:Absorption, Utilization and Functionality. *Food Sci.* **2013**, *34*, 375–391.

16. Yang, R.Y.; Zhang, Z.F.; Pei, X.Y.; Han, X.L.; Wang, J.B.; Wang, L.L.; Long, Z.; Shen, X.Y.; Li, Y.i. Immunomodulatory effects of marine oligopeptide preparation from Chum Salmon (Oncorhynchus keta) in mice. *Food Chem.* **2009**, *113*, 464–470. [CrossRef]

17. Gao, S.; Hong, H.; Zhang, C.Y.; Wang, K.; Zhang, B.H.; Han, Q.A.; Liu, H.G.; Luo, Y.K. Immunomodulatory effects of collagen hydrolysates from yak (Bos grunniens) bone on cyclophosphamide-induced immunosuppression in BALB/c mice. *J. Funct. Foods* **2019**, *60*, 103420. [CrossRef]

18. Li, W.; Ye, S.W.; Zhang, Z.W.; Tang, J.C.; Jin, H.X.; Huang, F.F.; Yang, Z.S.; Tang, Y.P.; Chen, Y.; Ding, G.F.; et al. Purification and Characterization of a Novel Pentadecapeptide from Protein Hydrolysates of Cyclina sinensis and Its Immunomodulatory Effects on RAW264.7 Cells. *Mar. Drugs* **2019**, *17*, 30. [CrossRef]

19. Yu, F.M.; Zhang, Z.W.; Ye, S.W.; Hong, X.X.; Jin, H.X.; Huang, F.F.; Yang, Z.S.; Tang, Y.P.; Chen, Y.; Ding, G.F. Immunoenhancement effects of pentadecapeptide derived from Cyclina sinensis on immune-deficient mice induced by Cyclophosphamide. *J. Funct. Foods.* **2019**, *60*, 103408. [CrossRef]

20. Huang, F.F.; Wang, J.J.; Yu, F.M.; Tang, Y.P.; Ding, G.F.; Yang, Z.S.; Sun, Y. Protective Effect of *Meretrix meretrix* Oligopeptides on High-Fat-Diet-Induced Non-Alcoholic Fatty Liver Disease in Mice. *Mar. Drugs* **2018**, *16*, 39. [CrossRef]

21. Ye, L.; Yan, J.; Zhang, W.; Zou, S.S.; Ye, S.W.; Yang, Z.S.; Yu, F.M.; Ding, G.F. Immunomodulatory effects of Meretrix meretrix oligopeptides on RAW264.7 cells. *J. Fish. China* **2019**, *43*, 24–32.

22. Kyung, D.; Sung, H.; Kim, Y. Global transcriptome analysis identifies weight regain-induced activation of adaptive immune responses in white adipose tissue of mice(Article). *Int. J. Obesity* **2018**, *42*, 755–764. [CrossRef] [PubMed]

23. Sadasivan, S.; Vasamsetti, B.; Singh, J.; Marikunte, V.; Oommen, A.; Jagannath, M.; Rao, R. Exogenous administration of spermine improves glucose utilization and decreases bodyweight in mice. *Eur. J. Pharmacol.* **2014**, *729*, 94–99. [CrossRef] [PubMed]

24. Kim, S.; Lee, S.; Lee, S. Rutecarpine ameliorates bodyweight gain through the inhibition of orexigenic neuropeptides NPY and AgRP in mice. *Biochem. Bioph. Res. Co.* **2009**, *389*, 437–442. [CrossRef]

25. Deng, Y.; Wang, L.; Wang, C. Tolerance-like innate immunity and spleen injury: A novel discovery via the weekly administrations and consecutive injections of PEGylated emulsions. *Int. J. Nanomed.* **2014**, *9*, 3645–3657. [CrossRef]

26. Mebius, R.E.; Kraal, G. Structure and function of the spleen. *Nat. Rev. Immunol.* **2005**, *5*, 606–616. [CrossRef]

27. Guo, L.; Sun, Y.; Wang, A. Effect of polysaccharides extract ofrhizoma atractylodis macrocephalaeon thymus, spleen and cardiac indexes, caspase-3 activity ratio, Smac/DIABLO and HtrA2/Omi protein and mRNA expression levels in aged rats. *Mol. Biol. Rep.* **2012**, *39*, 9285–9290. [CrossRef]

28. Zhe, W.; Yu, Z.; Yan, Y. Lipopolysaccharide Preconditioning Increased the Level of Regulatory B cells in the Spleen after Acute Ischaemia/Reperfusion in Mice. *Brain Res.* **2018**, *1701*, 46–57.

29. Salem, M.L.; Al-Khami, A.A.; El-Nagaar, S.A.; Zidan, A.A.; Al-Sharkwi, I.M.; Díaz-Montero, C.M.; Cole, D.J. Kinetics of rebounding of lymphoid and myeloid cells in mouse peripheral blood, spleen and bone marrow after treatment with cyclophosphamide. *Cell. Immunol.* **2012**, *276*, 67–74. [CrossRef]

30. Brian, J.; Jason, C.; John, C. Reduction of MHC-I expression limits T-lymphocyte-mediated killing of Cancer-initiating cells. *BMC Cancer* **2018**, *18*, 469.

31. Battista, J.; Tallmadge, R.; Stokol, T.; Felippe, M. Hematopoiesis in the equine fetal liver suggests immune preparedness. *Immunogenetics* **2014**, *66*, 635–649. [CrossRef] [PubMed]

32. Vidarsson, G.; Dekkers, G.; Rispens, T. IgG Subclasses and Allotypes: From Structure to Effector Functions. *Front. Immuno.* **2014**, *5*. [CrossRef] [PubMed]

33. Jiang, H.W.; Shen, X.H.; Chen, Z.Y.; Liu, F.; Wang, T.; Xie, Y.K.; Ma, C. Nociceptive neuronal Fc-gamma receptor I is involved in IgG immune complex induced pain in the rat. *Brain Behav. Immun.* **2017**, *62*, 351–361. [CrossRef] [PubMed]

34. Sangwan, V.; Tomar, S.K.; Ali, B.; Singh, R.R.B.; Singh, A.K. Galactooligosaccharides reduce infection caused by Listeria monocytogenes and modulate IgG and IgA levels in mice. *Int. Dairy J.* **2015**, *41*, 58–63. [CrossRef]

35. Pan, D.D.; Wu, Z.; Liu, J. Immunomodulatory and hypoallergenic properties of milk protein hydrolysates in ICR mice, Journal of Dairy Science. *J. Dairy Sci.* **2013**, *96*, 4958–4964. [CrossRef]

36. Chen, Y.; Tang, J.B.; Wang, X.K.; Sun, F.X.; Liang, S.J. An immunostimulatory polysaccharide (SCP-IIa) from the fruit of Schisandra chinensis (Turcz.). *Int. J. Biol. Macromol.* **2012**, *50*, 844–848. [CrossRef]

37. Liu, J.C.; Sun, H.; Nie, C.X.; Ge, W.X.; Wang, Y.Q.; Zhang, W.J. Oligopeptide derived from solid-state fermented cottonseed meal significantly affect the immunomodulatory in balb/c mice treated with cyclophosphamide. *Food Sci. Biotechnol.* **2018**, *27*, 1791–1799. [CrossRef]

38. Karayannopoulou, M.; Anagnostou, T.; Margariti, A. Evaluation of blood T-lymphocyte subpopulations involved in host cellular immunity in dogs with mammary cancer. *Vet. Immunol. Immunop.* **2017**, *186*, 45–50. [CrossRef]

39. Actor, J. T Lymphocytes: Ringleaders of Adaptive Immune Function. In *Introductory Immunology*, 1st ed.; Academic Press: Cambridge, MA, USA, 2014; pp. 42–58.

40. Nikš, M.; Otto, M.; Bušová, B.; Stefanovic, J. Quantification of proliferative and suppressive responses of human T lymphocytes following ConA stimulation. *J. Immunol. Methods* **1990**, *126*, 263–271. [CrossRef]

41. Mustafa, W.; Al-Saleem, F.H.; Nasser, Z. Immunization of mice with the non-toxic HC50 domain of botulinum neurotoxin presented by rabies virus particles induces a strong immune response affording protection against high-dose botulinum neurotoxin challenge. *Vaccine* **2011**, *29*, 4638–4645. [CrossRef]

42. Ye, F.; Yan, S.; Xu, L. Tr1 regulatory T cells induced by ConA pretreatment prevent mice from ConA-induced hepatitis. *Immunol. Lett.* **2009**, *122*, 198–207. [CrossRef] [PubMed]

43. Zhang, X. Immunization with Pseudomonas aeruginosa outer membrane vesicles stimulates protective immunity in mice. *Vaccine* **2018**, *36*, 1047–1054. [CrossRef] [PubMed]

44. Wang, S.; Huang, S.; Ye, Q.H.; Zeng, X.F.; Yu, H.T.; Qi, D.S.; Qiao, S.Y. Prevention of Cyclophosphamide-Induced Immunosuppression in Mice with the Antimicrobial Peptide Sublancin. *Clin. Dev. Immunol.* **2018**, *2018*, 1–11. [CrossRef] [PubMed]

45. Meng, Y.Y.; Li, B.L.; Jin, D.; Zhan, M.; Lu, J.J.; Huo, G.C. Immunomodulatory activity ofLactobacillus plantarumKLDS1.0318 in cyclophosphamide-treated mice. *Food Nutr. Res.* **2018**, *62*, 1296. [CrossRef]

46. Cai, G.D.; Sun, K.; Wang, T. Mechanism and effects of Zearalenone on mouse T lymphocytes activation in vitro. *Ecotox. Environ. Safe* **2018**, *162*, 208–217. [CrossRef]

47. Ye, Y.; Zhang, Y.; Lu, X.; Huang, X.Y.; Zeng, X.F.; Lai, X.Q.; Zeng, Y.Y. The anti-inflammatory effect of the SOCC blocker SK&F 96365 on mouse lymphocytes after stimulation by ConA or PMA/ionomycin. *Immunobiology* **2011**, *216*, 1044–1053.

 *molecules*

Article

Marine Collagen Peptides Promote Cell Proliferation of NIH-3T3 Fibroblasts via NF-κB Signaling Pathway

Fei Yang [1], Shujie Jin [2] and Yunping Tang [2,*] 

[1] School of Nursing, Zhejiang Chinese Medical University, Hangzhou 310053, China; yangfei19870826@126.com

[2] Zhejiang Provincial Engineering Technology Research Center of Marine Biomedical Products, School of Food and Pharmacy, Zhejiang Ocean University, Zhoushan 316022, China; m18868006087@163.com

* Correspondence: tangyunping1985@zjou.edu.cn; Tel.: +86-0580-229-9809; Fax: +86-0580-229-9866

Received: 30 September 2019; Accepted: 18 November 2019; Published: 19 November 2019

Abstract: Marine collagen peptides (MCPs) with the ability to promote cell proliferation and migration were obtained from the skin of *Nibea japonica*. The purpose of MCPs isolation was an attempt to convert the by-products of the marine product processing industry to high value-added items. MCPs were observed to contain many polypeptides with molecular weights ≤ 10 kDa and most amino acid residues were hydrophilic. MCPs (0.25–10 mg/mL) also exhibited 2, 2-diphenyl-1-picrylhydrazyl (DPPH), hydroxyl, superoxide anion, and 2′-azino-bis-3-ethylbenzothiazoline-6-sulfonic acid (ABTS) radical scavenging activities. Furthermore, MCPs promoted the proliferation of NIH-3T3 cells. In vitro scratch assays indicated that MCPs significantly enhanced the scratch closure rate and promoted the migration of NIH-3T3 cells. To further determine the signaling mechanism of MCPs, western blotting was used to study the expression levels of nuclear factor kappa-B (NF-κB) p65, IκB kinase α (IKKα), and IκB kinase β (IKKβ) proteins of the NF-κB signaling pathway. Our results indicated protein levels of NF-κB p65, IKKα and IKKβ increased in MCPs-treated NIH-3T3 cells. In addition, MCPs increased the expression of epidermal growth factor (EGF), fibroblast growth factor (FGF), vascular endothelial growth factor (VEGF), and transforming growth factor (TGF-β) in NIH-3T3 cells. Therefore, MCPs, a by-product of *N. japonica*, exhibited potential wound healing abilities in vitro.

Keywords: *Nibea japonica*; marine collagen peptides; proliferation; wound healing; processing by-products

1. Introduction

Owing to the boom in marine product processing industry, there is a huge production of by-products that are either discarded or simply used as animal feed or fertilizer [1,2]. Hence, there is an urgent requirement to explore methods for using such by-products to yield high value-added items. Bioactive peptides from marine resources possess several physiological functions, including antioxidative [3], anticancer [4], antibacterial [5], angiotensin-converting-enzyme (ACE) inhibitory [6], immunomodulatory [7], hepatoprotective [8], and wound healing activities [9]. Therefore, extraction of bioactive peptides from marine wastes and by-products might offer new avenues for their utilization, consequently preventing environmental pollution and creating enormous economic benefits [6,10].

Collagen extracted from marine by-products is in demand due to the absence of religious restrictions on its use, low immunogenicity, and non-cytotoxicity [11,12]. Marine collagen undergoes enzymatic and chemical hydrolysis to generate marine collagen peptides (MCPs). [13]. Compared to collagen, MCPs possess several advantages, such as ease of absorption for its lower molecular weight and unique physiological functions (including antioxidation) [3,14], high affinity to calcium [15], antihypertensive [16,17], and wound healing activities [18,19].

Currently, the effects of collagen peptides or their combinations with other functional ingredients on wound healing have been the focus of many studies because of their outstanding antioxidant and antimicrobial properties [13,18]. Wound healing is a complex process involving cell-matrix interactions, inflammation, new tissue formation and tissue remodeling [19,20]. Fibroblasts are responsible for regeneration and remodeling of connective tissue in healthy skin [18,20]. So far, there are only few studies dedicated to explore the mechanism of wound healing induced by MCPs in vitro or in vivo. Previous researches have shown a close association between wound healing and nuclear factor kappa enhancer binding protein (NF-κB) signaling pathway [5,21]. NF-kB is related to cell proliferation, cell adhesion, inflammation and elimination of reactive oxygen species (ROS) [20]. In addition, the NF-kB signaling pathway has been reported to be involved with cutaneous [7,22] and corneal epithelial wound healing [21]. As NF-kB signaling and MCPs are both connected to wound healing, we hypothesized that this pathway might play an important role in wound healing induced by MCPs.

Previously, marine collagen was successfully obtained from the skin of *Nibea japonica*, and the physicochemical properties and biocompatibility were determined [11,23]. In an attempt to identify functional MCPs, in the present study, MCPs were extracted from *N. japonica* skins for their functional assessment. We also analyzed the molecular weight, amino acid content and antioxidant activities of the extracted MCPs. Our study showed that MCPs can promote cell proliferation and migration of NIH-3T3 fibroblasts via the NF-kB signaling pathway.

2. Results and Discussions

2.1. Determination of Molecular Weight Distribution of MCPs

The HPLC spectrum of the standard molecular weight samples is shown in Figure 1A and the regression equation obtained is as follows:

$$\text{lgMw} = -0.2753\text{Rt} + 7.3148 \tag{1}$$

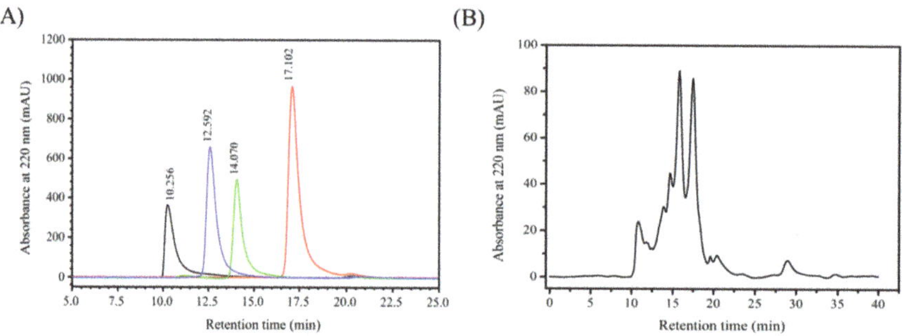

Figure 1. The HPLC spectra of the standard molecular weight samples (**A**) and marine collagen peptides (MCPs) from skin of *Nibea japonica* (**B**).

The coefficient of regression (R^2) was 0.9652 indicating good linear relationship and the molecular weight distribution of MCPs could be determined based on the above equation. The HPLC spectrum of MCPs from the *N. japonica* skins is shown in Figure 1B. Components less than 1, 3, 5 and 10 kDa accounted for 55.25%, 79.29%, 85.71% and 90.31% of the spectrum respectively indicating these MCPs primarily contained a large number of low molecular weight polypeptides. Furthermore, *N. japonica* MCPs had better water solubility than the marine collagen [11] essentially because their low molecular weight structures possess many water-exposed polar amino acid residues, leading to the formation of more hydrogen bonds [24,25].

2.2. Amino Acid Content of MCPs

The amino acid content of MCPs from *N. japonica* skins is shown in Figure 2. The studied MCPs comprised seven essential amino acids (11.49%) and ten non-essential amino acids (70.48%). Glycine was the principal amino acid in MCPs, accounting for approximately 21.22% of the total amino acid composition, followed by proline (10.55%), alanine (9.79%), hydroxyproline (9.28%), arginine (7.47%) and glutamic acid (4.48%). Furthermore, no cysteine was detected in the concerned MCPs. In our previous studies, we confirmed that the collagen from *N. japonica* skins is a type I collagen [11,23]. Cysteine being exclusively present in type III collagen [12] our results confirm our previous observation. MCPs usually contain a high concentration of Gly-Xaa-Yaa triplets, where Xaa is usually proline and Yaa is most likely hydroxyproline [11,12]. The high content of glycine, proline and hydroxyproline in MCPs was consistent with the high frequency of occurrence of the Gly-Pro-Hyp sequence in the collagen. Furthermore, the glycine (21.22%), proline (10.55%), hydroxyproline (9.28%) and arginine (7.47%) contents in MCPs from *N. japonica* skins was similar to that in MCPs from tilapia skin (where the percentage of glycine, proline, hydroxyproline, and arginine were 20.92%, 11.32%, 10.28% and 7.96%, respectively) [13]. The majority of the amino acid residues were hydrophilic, such as hydroxyproline, arginine, glutamic acid, and aspartic acid. This was consistent with the good water solubility of MCPs from *N. japonica* skins.

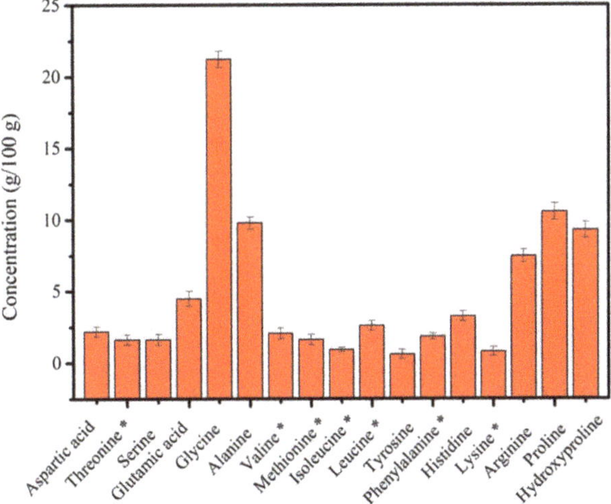

Figure 2. Amino acids content of MCPs extracted from skin of *Nibea japonica*. Note: * essential amino acid. All assays were performed in triplicate.

2.3. Antioxidant Activity of MCPs

NADPH oxidase stimulates inflammatory cells to produce large amounts of ROS during the inflammatory phase of wound healing [26,27]. Normally, ROS are scavenged by antioxidants, and there is a balance between ROS production and neutralization [26]. On the contrary, this balance is disturbed in a wound, where excessive ROS are produced. Excessive ROS induction is associated with activation of pro-apoptotic proteins resulting in cell death and necrosis and can be harmful for wound healing. MCPs are widely used in the skin care industry to promote wound healing due to their antioxidant properties and other beneficial properties [28]. Therefore, the antioxidant activities of MCPs (0.25–10 mg/mL) from the skin of *N. japonica* were evaluated using four different radical scavenging assays.

As illustrated in Figure 3, MCPs (0.25–10 mg/mL) obtained from *N. japonica* skins could scavenge DPPH, hydroxyl, superoxide anion and ABTS radicals. The concentration of MCPs was related to the scavenging activities of these four free radicals. The scavenging activities of these four free radicals also increased in proportion to MCPs concentration. However, as shown in Figure 3, the antioxidant activities of MCPs were relatively lower than that of ascorbic acid (approximately 0–60% at concentrations between 0.25 and 10 mg/mL), and should be improved for use in wound healing. Recently, several functional ingredients, such as chitosan, chemically modified chitosan or nicotinamide were used to improve the antioxidant activity of peptides to promote wound healing. For example, the *N*-succinyl chitosan-collagen peptide copolymer manufactured with transglutaminase possessed better antioxidant activity and could be used as a wound healing biomaterial [28]. Nicotinyl-isoleucine-valine-histidine (NA-IVH), manufactured by combining nicotinamide and jellyfish peptides (IVH), showed significant enhancement of radical scavenging function and can promote wound healing under hyperglycemic condition [26]. Therefore, MCPs have to be modified to improve their wound healing properties for subsequent application in wound healing.

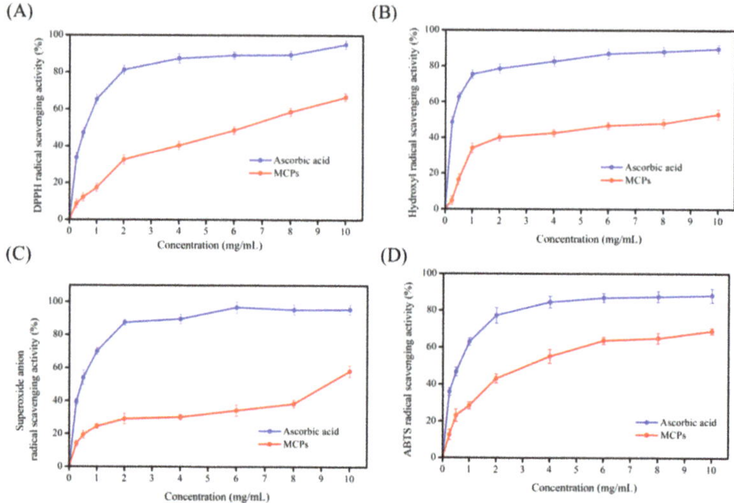

Figure 3. 2, 2-diphenyl-1-picrylhydrazyl (DPPH) (**A**), Hydroxyl (**B**), superoxide anion (**C**) and ABTS (**D**) radical scavenging activities of MCP s from *Nibea japonica* skins. All assays were performed in triplicate.

2.4. Cell Proliferation of NIH-3T3

Various types of cells are known to undergo migration and proliferation during wound healing. Fibroblasts are the key components of normal wound healing and play an important role from late inflammation to complete epithelialization [29]. The present study demonstrated that MCPs have the potential to promote the growth of NIH-3T3 cells. As shown in Figure 4, the viability rate of NIH-3T3 cells treated with varied concentrations of MCPs increased significantly post 72 h of incubation. The viability of cells treated with 25 µg/mL MCPs was 37% more than that of the negative control (NC) group, but was lower than that of the positive control (PC) group. Our observation is in agreement with the results obtained using MCPs from tilapia, which promoted L929 fibroblast proliferation [18]. Thus, MCPs showed significant proliferation in vitro and have potential to be used for wound healing or cosmetic application.

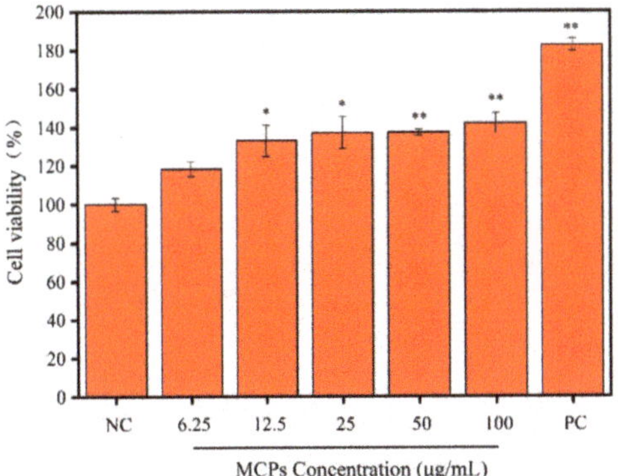

Figure 4. Relative cell viability as affected by 72 h treatment of different concentrations of MCPs from *Nibea japonica* skin. Negative control (NC): adding 0.4% serum DMEM to cells; Experimental group: MCP was dissolved by 0.4% serum DMEM in concentrations of 6.25, 12.5, 25, 50, and 100 µg/mL and then added to the prepared cells; Positive control (PC): adding 10% serum DMEM to cells. * $p < 0.05$ and ** $p < 0.001$ vs. NC. The data were expressed as the mean ± standard deviation ($x \pm s$, $n = 6$).

2.5. Effect of MCPs on the Scratch Wound Closure In Vitro

Fibroblast migration can accelerate the process of wound re-epithelialization and promote wound closure during healing [30]. Previously, in vitro scratch test has often been used to simulate wound healing [18,20]. Therefore, we used the above assay on NIH-3T3 cells to evaluate the effect of MCPs from *N. japonica* skins on the wound healing process. As shown in Figure 5, the migration of cells to the scratched area was evident after 12 h and 24 h. In addition, in the presence of MCPs, the wound area was significantly reduced in a dose-dependent manner compared to the control group without MCPs. Significant scratch closure mediated by MCPs was observed after 24 h. In particular, the effect of 50 µg/mL MCPs on in vitro wound healing was highly statistically significant (Figure 5B) and the scratch was almost completely sealed (Figure 5A). Our results indicated that MCPs was capable of inducing NIH-3T3 cell migration and potentially promote wound healing. This may be because abundant amino acids residues in MCPs provide a suitable environment for NIH-3T3 cells to proliferate and migrate (although the mechanism is not clearly delineated).

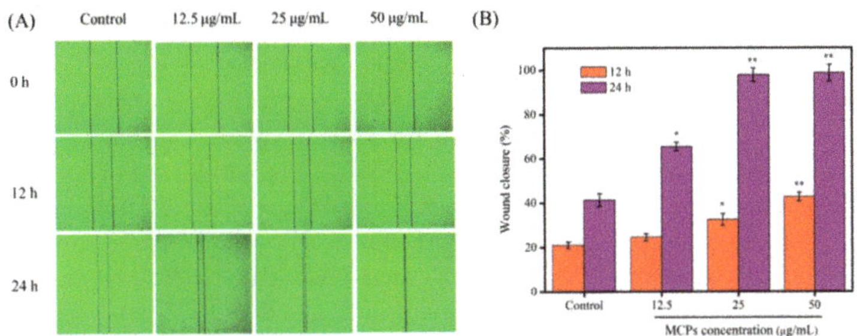

Figure 5. Effect of MCPs from *Nibea japonica* skins on the scratch closure in vitro. (**A**) Representative optical images showed the cells migrated toward wound gap after 12 h and 24 h incubation; (**B**) Wound closure rate (%) that affected by MCPs for 12 h and 24 h. The data was obtained by using Image J 1.38 software and were expressed as the mean ± standard deviation ($x \pm s$, $n = 6$). * $p < 0.05$ and ** $p < 0.001$ vs. control.

2.6. MCPs Activated the NF-κB Signaling Pathway in NIH-3T3 Fibroblasts

NF-κB is a transcription factor that regulates the expression of multiple genes involved in a variety of cellular functions including cell migration, proliferation, adhesion and survival [20,31]. Therefore, for further confirmation of role of MCPs towards activation of the signaling pathway through NF-κB the protein expression levels of some related proteins were evaluated using western blotting. As shown in Figure 6, NF-κB p65, IκB kinase α (IKKα), and IκB kinase β (IKKβ) levels increased significantly after treatment of different concentrations of MCPs in a dose-dependent manner. These results indicated that MCPs can promote NIH-3T3 cell migration and proliferation via the NF-κB signaling pathway.

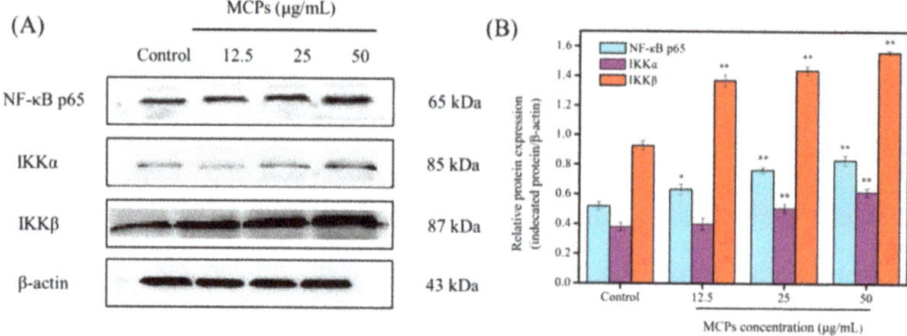

Figure 6. MCPs activated the NF-κB signaling pathway and increased expression of its target pathways in NIH-3T3 fibroblast cells. (**A**) Western blot analysis of the NF-κB p65, IKKα, and IKKβ in the NIH-3T3 cells treated with different concentrations MCPs overnight. (**B**) The expression levels of NF-κB p65, IKKα, and IKKβ analyzed by western blotting. The data was obtained by using Image J 1.38 software and were expressed as the mean ± standard deviation ($x \pm s$, $n = 6$). * $p < 0.05$ and ** $p < 0.001$ vs. control.

2.7. Western Blot Analysis of Growth Factors

Wound healing is a complex process regulated by different signaling pathways, various cytokines and certain growth factors [19,32]. Epidermal growth factor (EGF) can enhance the migration and proliferation of fibroblasts. In addition, EGF promotes angiogenesis and epithelization and triggers growth factor secretion by fibroblasts, which ultimately leads to accelerated wound healing [33]. Fibroblast growth factor (FGF) can enhance angiogenesis, cell migration and proliferation to promote wound healing [34]. Vascular endothelial growth factor (VEGF) is the main growth factor that triggers angiogenesis and stimulates wound healing [20,32]. Transforming growth factor (TGF-β) can also induce various processes such as secretion of extracellular matrix proteins, proliferation, migration, and angiogenesis [35]. Therefore, we further assessed the effects of MCPs from *N. japonica* skin on the expression of EGF, FGF, VEGF, and TGF-β in this study. As shown in Figure 7, the protein levels of EGF, FGF, VEGF, and TGF-β increased significantly after treatment with various concentrations of MCPs. These results support the notion that MCPs can be applied for promoting wound healing. In addition, a continuous over-expression of such growth factors, without a turning-back point towards their initial levels after wound healing, may be linked to other non-beneficial proliferation-related manifestations such as cancerous neoangiogenesis because of an unresolved inflammatory process [36]. So, further in vivo experiments should be applied to the wound surface of the skin to demonstrate the effect of MCPs, and the role of NF-κB signaling pathway or growth factors in promoting wound healing.

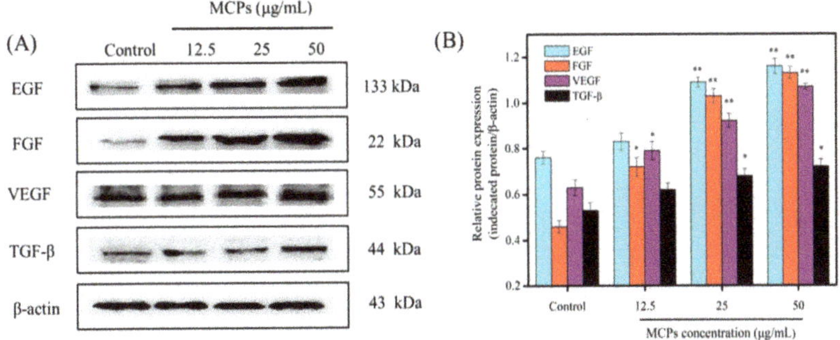

Figure 7. Effect of MCPs on the protein expression levels of epidermal growth factor (EGF), fibroblast growth factor (FGF), vascular endothelial growth factor (VEGF), and transforming growth factor (TGF-β) in NIH-3T3 fibroblast cells. (**A**) Western blot analysis of the EGF, FGF, VEGF, and TGF in the NIH-3T3 cells treated with different concentrations MCPs overnight (**B**) Protein expression levels of EGF, FGF, VEGF, and TGF analyzed by western blotting. The data was obtained by using Image J 1.38 software and were expressed as the mean ± standard deviation ($x ± s$, $n = 6$). * $p < 0.05$ and ** $p < 0.001$ vs. control.

3. Materials and Methods

3.1. Materials

N. japonica skins available in our laboratory [11], and the NIH-3T3 fibroblasts were purchased from the Cell Bank of Chinese Academy of Sciences (Shanghai, China). Antibodies raised against NF-κB p65 (cat. No. AF0246), IKKα (cat. No. AF0198), IKKβ (cat. No. AI137), VEGF (cat. no. AF1309), and TGF-β (cat. No. AF0198) were purchased from Beyotime Biotechnology (Shanghai, China). β-actin (cat. no. K200058M) was purchased from Solarbio (Beijing, China). Detection antibodies for EGF (cat. no. 184265) and FGF (cat. no. ab171941) were procured from Abcam (Cambridge, England). MTT cell proliferation and cytotoxicity assay kit (AR1156) was purchased from Boster Biological Technology co. ltd (Wuhan, China). All other reagents were of analytical grade.

3.2. Preparation of MCPs from N. japonica skin

The non-collagenous proteins and fat were removed from the *N. japonica* skins following protocol described by Tang et al. [11]. Following it, the fish skins were heated at 100 °C for 10 min, and hydrolyzed in presence of neutral protease (1500 U/g). We adjusted the initial pH of the solution to 7.0 and the enzymatic hydrolysis was performed at 45 °C for 3 h. The above step was followed by enzyme deactivation at 100 °C for 10 min. After centrifugation, the supernatant of MCPs was collected and lyophilized for further study.

3.3. Determination of the Molecular Weight Distribution of MCPs

The molecular weight distribution of MCPs was analyzed using high pressure liquid chromatography (HPLC) (Agilent 1200, CA, USA). We used a TSK gel G2000 SWXL analytical column (4.6 × 250 mm, 5 μm) at UV 220 nm at 25 °C, and a mobile phase of acetonitrile/water/trifluoroacetic acid (45:55:0.1) at the flow rate of 0.5 mL/min. The standard samples comprised of peroxidase (40,000 Da), aprotinin (6500 Da), Arg-Val-Ala-Pro-Glu-Glu-His-Pro-Val-Glu-Gly-Arg-Tyr-Leu-Val (1750 Da) [7], and Tyr-Val-Pro-Gly-Pro (530 Da) [4] which were loaded into the column by turn. The standard curve of retention time and absorbance was plotted. The MCPs solution was then filtered using 0.22 μm micropore film and injected under the same conditions. Finally, the molecular weight distribution of MCPs was calculated according to the standard curve equation.

3.4. Amino Acid Content

The amino acid content was determined according to the Chinese national standard (GB5009124-2016). MCPs were first dissolved in 6 M HCl solution and hydrolyzed at 110 °C for 24 h. The hydrolysate was further diluted with citric acid buffer and analyzed using an amino acid analyzer L-8900 (Hitachi, Tokyo, Japan). The hydroxyproline content was analyzed as per procedure described by Tang et al. [11].

3.5. Antioxidant Activity of MCPs

The hydroxyl, 2, 2-diphenyl-1-picrylhydrazyl (DPPH), 2, 2′-azino-bis-3-ethylbenzothiazoline-6-sulfonic acid (ABTS), and superoxide anion radical scavenging activity of MCPs was analyzed according to Zhao et al. [37].

3.6. Proliferation of NIH-3T3 Fibroblasts in Presence of MCPs

The MTT assay was employed to assess proliferation of MCPs [11]. Briefly, NIH-3T3 cells were seeded in a 96-well plate at a density of 2×10^5 cells/mL and cultured in complete medium (Dulbecco's modified Eagle's medium (DMEM) supplemented with 10% fetal bovine serum (FBS), 100 U/mL penicillin and 100 µg/mL streptomycin) at 37 °C in a 5% CO_2 incubator. Upon 80–90% cellular confluence the culture medium was substituted with fresh maintenance medium (DMEM contains 0.4% FBS), and the cells were further grown for duration of 24 h at 37 °C. Further the cells were exposed to different concentrations of MCPs (0, 6.25, 12.5, 25, 50 and 100 µg/mL) in the maintenance medium, and cultured for 24, 48 and 72 h respectively. Cells grown in the same volume of complete medium served as the positive control group. The optical density (OD) at 490 nm was determined using a microplate reader (SpectraMa, Molecular Devices Co., San Jose, CA, USA), and relative cell viability (%) was calculated using the following formula:

$$\text{Relative cell viability (\%)} = [1 - (\text{OD treated}/\text{OD untreated})] \times 100\% \tag{2}$$

3.7. In Vitro Scratch Wound Assay

NTH-3T3 cells were seeded in a 6-well plate at a density of 2×10^5 cells/mL and incubated in complete medium until cell confluence reached about 80–90%. The cells were further grown for next 24 h at 37 °C in 5% CO_2 incubator. A uniform scratch wound was created using a 200 µL sterile pipette tip, and the wound debris was removed through phosphate buffer saline (PBS) wash. The scratched cells were then treated with different concentrations of MCPs (0, 12.5, 25 and 50 µg/mL) and cultured for 12 or 24 h. Scratch closure was evaluated using an inverted microscope (Olympus, Tokyo, Japan) and the scratch area was analyzed using the Image J 1.38 software (NIH, Bethesda, MD, USA). We enumerated the scratch closure rate (%) based on the following formula:

$$\text{Scratch closure rate (\%)} = (A_0 - A_t)/A_0 \times 100\% \tag{3}$$

where A_0 represents the scratch area at 0 h and A_t represents the same at the designated time point.

3.8. Western Blot Analysis

Western blotting of target proteins helped to confirm the proliferation of MCPs on NIH-3T3 cells., We used the technique according to Jiang et al. [38] with certain modifications. We used a seeding density of 2×10^5 cells/mL for culturing NTH-3T3 cells and treated with varied concentrations of MCPs (0, 12.5, 25 and 50 µg/mL) for 24 h. Subsequently, the cells were collected and lysed in radioimmunoprecipitation assay (RIPA) lysis solution. Protein concentration of cellular lysates was obtained using the bicinchoninic acid (BCA) protein assay. Further, an equivalent amount of denatured protein sample (30 µg) was resolved using 12% sodium dodecyl sulfate (SDS)-polacrylamide gel. After electrophoresis, the gel was transferred onto a polyvinylidene difluoride (PVDF) membrane. Non-specific binding was prevented through incubation with 5% skimmed milk for 1 h followed by overnight incubation with diluted (1:1000) primary antibodies (NF-κB p65, IKKα, IKKβ, VEGF, EGF, FGF, and TGF-β) at 4 °C. We finally incubated the membranes in diluted (1:1000) secondary antibodies for 1 h at room temperature. The target protein bands were visualized using enhanced chemiluminescence and the density was enumerated using the Image J 1.38 software (NIH, Bethesda, MD, USA). We used β-Actin as an internal control.

3.9. Statistical Analysis

We represented all experimental data as the mean ± standard deviation ($x \pm s$, $n = 6$) and analyzed using the SPSS software version 24.0 (SPSS Inc., Chicago, IL, USA). Statistical significance of the data was determined using one-way analysis of variance (ANOVA).

4. Conclusions

In the present study, MCPs prepared from the skin of *N. japonica* exhibited potential cell proliferation and migration activities. Our results indicated that MCPs are rich in polypeptides with molecular weights ≤ 10 kDa. MCPs could scavenge DPPH, hydroxyl, superoxide anion, and ABTS radical as well as promoted the proliferation and migration of NIH-3T3 cells. In vitro scratch assays also reflected that MCPs significantly affect the scratch closure rate. MCPs further increased the protein levels of NF-κB p65, IKKα, and IKKβ, which are prominent members of the NF-κB signaling pathway, as well as those of certain growth factors such as EGF, FGF, VEGF, and TGF-β in NIH-3T3 cells (as revealed through western blotting). In conclusion, our results indicated that MCPs from the skin of *N. japonica* possess potential to promote wound healing. The findings may provide guidance for high value-added utilization by-products of marine processing industry. In the future, in vivo experiments are needed to apply to the wound surface of the skin to demonstrate role of MCPs in promoting wound healing.

Author Contributions: Y.T. conceived and designed the experiments. F.Y. and S.J. performed the experiments and carried out statistical analysis of the data. F.Y. and S.J. wrote the manuscript.

Funding: This work was financially supported by the National Natural Science Foundation of China (grant No. 41806153).

Conflicts of Interest: The authors declare no conflict of interest.

References

1. Lapena, D.; Vuoristo, K.S.; Kosa, G.; Horn, S.J.; Eijsink, V.G.H. Comparative Assessment of Enzymatic Hydrolysis for Valorization of Different Protein-Rich Industrial Byproducts. *J. Agric. Food Chem.* **2018**, *66*, 9738–9749. [CrossRef]
2. Shavandi, A.; Hou, Y.; Carne, A.; McConnell, M.; Bekhit, A.E.-D.A. Marine Waste Utilization as a Source of Functional and Health Compounds. *Adv. Food Nutr. Res.* **2019**, *87*, 187–254.
3. Wang, B.; Wang, Y.M.; Chi, C.F.; Luo, H.Y.; Deng, S.G.; Ma, J.Y. Isolation and characterization of collagen and antioxidant collagen peptides from scales of croceine croaker (*Pseudosciaena crocea*). *Mar. Drugs* **2013**, *11*, 4641–4661. [CrossRef]
4. Li, X.J.; Tang, Y.P.; Yu, F.M.; Sun, Y.; Huang, F.F.; Chen, Y.; Yang, Z.S.; Ding, G.F. Inhibition of Prostate Cancer DU-145 Cells Proliferation by *Anthopleura anjunae* Oligopeptide (YVPGP) via PI3K/AKT/mTOR Signaling Pathway. *Mar. Drugs* **2018**, *16*, 16. [CrossRef]
5. Zhou, Q.J.; Wang, J.; Mao, Y.; Liu, M.; Su, Y.Q.; Ke, Q.Z.; Chen, J.; Zheng, W.Q. Molecular structure, expression and antibacterial characterization of a novel antimicrobial peptide NK-lysin from the large yellow croaker *Larimichthys crocea*. *Aquaculture* **2019**, *500*, 315–321. [CrossRef]
6. Yu, F.M.; Zhang, Z.W.; Luo, L.W.; Zhu, J.X.; Huang, F.F.; Yang, Z.S.; Tang, Y.P.; Ding, G.F. Identification and Molecular Docking Study of a Novel Angiotensin-I Converting Enzyme Inhibitory Peptide Derived from Enzymatic Hydrolysates of Cyclina sinensis. *Mar. Drugs* **2018**, *16*, 411. [CrossRef]

7. Li, W.; Ye, S.; Zhang, Z.; Tang, J.; Jin, H.; Huang, F.; Yang, Z.; Tang, Y.; Chen, Y.; Ding, G.; et al. Purification and Characterization of a Novel Pentadecapeptide from Protein Hydrolysates of *Cyclina sinensis* and Its Immunomodulatory Effects on RAW264.7 Cells. *Mar. Drugs* **2019**, *17*, 30. [CrossRef]

8. Nouvong, A.; Ambrus, A.M.; Zhang, E.R.; Hultman, L.; Coller, H.A. Reactive oxygen species and bacterial biofilms in diabetic wound healing. *Physiol. Genomics* **2016**, *48*, 889–896. [CrossRef]

9. Xu, C.; Zhang, R.; Wen, Z.Y. Bioactive compounds and biological functions of sea cucumbers as potential functional foods. *J. Funct. Foods* **2018**, *49*, 73–84. [CrossRef]

10. Umnyakova, E.S.; Gorbunov, N.P.; Zhakhov, A.V.; Krenev, I.A.; Ovchinnikova, T.V.; Kokryakov, V.N.; Berlov, M.N. Modulation of Human Complement System by Antimicrobial Peptide Arenicin-1 from Arenicola marina. *Mar. Drugs* **2018**, *16*, 480. [CrossRef]

11. Tang, Y.; Jin, S.; Li, X.; Li, X.; Hu, X.; Chen, Y.; Huang, F.; Yang, Z.; Yu, F.; Ding, G. Physicochemical Properties and Biocompatibility Evaluation of Collagen from the Skin of Giant Croaker (*Nibea japonica*). *Mar. Drugs* **2018**, *16*, 222. [CrossRef]

12. Khong, N.M.H.; Yusoff, F.M.; Jamilah, B.; Basri, M.; Maznah, I.; Chan, K.W.; Armania, N.; Nishikawa, J. Improved collagen extraction from jellyfish (*Acromitus hardenbergi*) with increased physical-induced solubilization processes. *Food Chem.* **2018**, *251*, 41–50. [CrossRef]

13. Hu, Z.; Yang, P.; Zhou, C.; Li, S.; Hong, P. Marine Collagen Peptides from the Skin of Nile Tilapia (*Oreochromis niloticus*): Characterization and Wound Healing Evaluation. *Mar. Drugs* **2017**, *15*, 102. [CrossRef]

14. You, L.; Zhao, M.; Regenstein, J.M.; Ren, J. Purification and identification of antioxidative peptides from loach (*Misgurnus anguillicaudatus*) protein hydrolysate by consecutive chromatography and electrospray ionization-mass spectrometry. *Food Res. Int.* **2010**, *43*, 1167–1173. [CrossRef]

15. Peng, Z.; Hou, H.; Zhang, K.; Li, B.F. Effect of calcium-binding peptide from Pacific cod (*Gadus macrocephalus*) bone on calcium bioavailability in rats. *Food Chem.* **2017**, *221*, 373–378. [CrossRef]

16. Byun, H.G.; Kim, S.K. Purification and characterization of angiotensin I converting enzyme (ACE) inhibitory peptides from Alaska pollack (*Theragra chalcogramma*) skin. *Process. Biochem.* **2001**, *36*, 1155–1162. [CrossRef]

17. Felician, F.F.; Xia, C.L.; Qi, W.Y.; Xu, H.M. Collagen from Marine Biological Sources and Medical Applications. *Chem. Biodivers.* **2018**, *15*, 1700557. [CrossRef]

18. Ouyang, Q.Q.; Hu, Z.; Lin, Z.P.; Quan, W.Y.; Deng, Y.F.; Li, S.D.; Li, P.W.; Chen, Y. Chitosan hydrogel in combination with marine peptides from tilapia for burns healing. *Int. J. Biol. Macromol.* **2018**, *112*, 1191–1198. [CrossRef]

19. Tingting, Y.; Kai, Z.; Bafang, L.; Hu, H. Effects of oral administration of peptides with low molecular weight from Alaska Pollock (*Theragra chalcogramma*) on cutaneous wound healing. *J. Funct. Foods* **2018**, *48*, 682–691.

20. Park, Y.R.; Sultan, M.T.; Park, H.J.; Lee, J.M.; Ju, H.W.; Lee, O.J.; Lee, D.J.; Kaplan, D.L.; Park, C.H. NF-kappa B signaling is key in the wound healing processes of silk fibroin. *Acta Biomater.* **2018**, *67*, 183–195. [CrossRef]

21. Wang, L.; Wu, X.; Shi, T.; Lu, L. Epidermal growth factor (EGF)-induced corneal epithelial wound healing through nuclear factor kappaB subtype-regulated CCCTC binding factor (CTCF) activation. *J. Biol. Chem.* **2013**, *288*, 24363–24371. [CrossRef] [PubMed]

22. Li, A.; Ye, L.; Yang, X.; Wang, B.; Yang, C.; Gu, J.; Yu, H. Reconstruction of the Catalytic Pocket and Enzyme-Substrate Interactions To Enhance the Catalytic Efficiency of a Short-Chain Dehydrogenase/Reductase. *Chemcatchem* **2016**, *8*, 3229–3233. [CrossRef]

23. Yu, F.; Zong, C.; Jin, S.; Zheng, J.; Chen, N.; Huang, J.; Chen, Y.; Huang, F.; Yang, Z.; Tang, Y.; et al. Optimization of Extraction Conditions and Characterization of Pepsin-Solubilised Collagen from Skin of Giant Croaker (*Nibea japonica*). *Mar. Drugs* **2018**, *16*, 29. [CrossRef] [PubMed]

24. Noman, A.; Xu, Y.S.; Al-Bukhaiti, W.Q.; Abed, S.M.; Ali, A.H.; Ramadhan, A.H.; Xia, W.S. Influence of enzymatic hydrolysis conditions on the degree of hydrolysis and functional properties of protein hydrolysate obtained from Chinese sturgeon (*Acipenser sinensis*) by using papain enzyme. *Process. Biochem.* **2018**, *67*, 19–28. [CrossRef]

25. Gbogouri, G.A.; Linder, M.; Fanni, J.; Parmentier, M. Influence of hydrolysis degree on the functional properties of salmon byproducts hydrolysates. *J. Food Sci.* **2004**, *69*, 615–622. [CrossRef]

26. Son, D.; Yang, D.; Sun, J.; Kim, S.; Kang, N.; Kang, J.; Choi, Y.-H.; Lee, J.; Moh, S.; Shin, D.; et al. A Novel Peptide, Nicotinyl-Isoleucine-Valine-Histidine (NA-IVH), Promotes Antioxidant Gene Expression and Wound Healing in HaCaT Cells. *Marine Drugs* **2018**, *16*, 262. [CrossRef]

27. Schafer, M.; Werner, S. Oxidative stress in normal and impaired wound repair. *Pharmacol. Res.* **2008**, *58*, 165–171. [CrossRef]

28. Pozzolini, M.; Millo, E.; Oliveri, C.; Mirata, S.; Salis, A.; Damonte, G.; Arkel, M.; Scarfi, S. Elicited ROS scavenging activity, photoprotective, and wound-healing properties of collagen-derived peptides from the marine sponge *Chondrosia reniformis*. *Mar. Drugs* **2018**, *16*, 465. [CrossRef]

29. Eming, S.A.; Martin, P.; Tomic-Canic, M. Wound repair and regeneration: Mechanisms, signaling, and translation. *Sci. Transl. Med.* **2014**, *6*, 16. [CrossRef]

30. Liu, H.; Mu, L.X.; Tang, J.; Shen, C.B.; Gao, C.; Rong, M.Q.; Zhang, Z.Y.; Liu, J.; Wu, X.Y.; Yu, H.N.; et al. A potential wound healing-promoting peptide from frog skin. *Int. J. Biochem. Cell Biol.* **2014**, *49*, 32–41. [CrossRef]

31. Chen, J.; Chen, Y.; Chen, Y.; Yang, Z.; You, B.; Ruan, Y.C.; Peng, Y. Epidermal CFTR Suppresses MAPK/NF-kappaB to Promote Cutaneous Wound Healing. *Cell Physiol. Biochem.* **2016**, *39*, 2262–2274. [CrossRef] [PubMed]

32. Xie, H.; Chen, X.; Shen, X.; He, Y.; Chen, W.; Luo, Q.; Ge, W.; Yuan, W.; Tang, X.; Hou, D.; et al. Preparation of chitosan-collagen-alginate composite dressing and its promoting effects on wound healing. *Int. J. Biol. Macromol.* **2018**, *107*, 93–104. [CrossRef] [PubMed]

33. Kuroyanagi, M.; Yamamoto, A.; Shimizu, N.; Ishihara, E.; Ohno, H.; Takeda, A.; Kuroyanagi, Y. Development of cultured dermal substitute composed of hyaluronic acid and collagen spongy sheet containing fibroblasts and epidermal growth factor. *J. Biomater. Sci. Polym. Ed.* **2014**, *25*, 1133–1143. [CrossRef] [PubMed]

34. Yoo, Y.; Hyun, H.; Yoon, S.J.; Kim, S.Y.; Lee, D.W.; Um, S.; Hong, S.O.; Yang, D.H. Visible light-cured glycol chitosan hydrogel dressing containing endothelial growth factor and basic fibroblast growth factor accelerates wound healing in vivo. *J. Ind. Eng. Chem.* **2018**, *67*, 365–372. [CrossRef]

35. Ning, J.L.; Zhao, H.L.; Chen, B.; Mi, E.Z.L.; Yang, Z.; Qing, W.H.; Lam, K.W.J.; Yi, B.; Chen, Q.; Gu, J.T.; et al. Argon Mitigates Impaired Wound Healing Process and Enhances Wound Healing In Vitro and In Vivo. *Theranostics* **2019**, *9*, 477–490. [CrossRef]

36. Tsoupras, A.B.; Iatrou, C.; Frangia, C.; Demopoulos, C.A. The implication of platelet activating factor in cancer growth and metastasis: Potent beneficial role of PAF-inhibitors and antioxidants. *Infect. Disord. Drug Targets* **2009**, *9*, 390–399. [CrossRef]

37. Zhao, W.H.; Chi, C.F.; Zhao, Y.Q.; Wang, B. Preparation, Physicochemical and Antioxidant Properties of Acid- and Pepsin-Soluble Collagens from the Swim Bladders of Miiuy Croaker (*Miichthys miiuy*). *Mar. Drugs* **2018**, *16*, 161. [CrossRef]

38. Jiang, S.; Jia, Y.; Tang, Y.; Zheng, D.; Han, X.; Yu, F.; Chen, Y.; Huang, F.; Yang, Z.; Ding, G. Anti-Proliferation Activity of a Decapeptide from Perinereies aibuhitensis toward Human Lung Cancer H1299 Cells. *Mar. Drugs* **2019**, *17*, 122. [CrossRef]

Sample Availability: Samples are available from the first or corresponding author.

 *molecules*

Review

An Updated Review on Pharmaceutical Properties of Gamma-Aminobutyric Acid

Dai-Hung Ngo [1] and **Thanh Sang Vo** [2,*]

[1]   Faculty of Natural Sciences, Thu Dau Mot University, Thu Dau Mot City 820000, Vietnam
[2]   NTT Hi-Tech Institute, Nguyen Tat Thanh University, Ho Chi Minh City 700000, Vietnam
*   Correspondence: vtsang@ntt.edu.vn; Tel.: +84-28-6271-7296

Academic Editors: María Dolores Torres and Elena Falqué López
Received: 27 May 2019; Accepted: 19 July 2019; Published: 24 July 2019

Abstract: Gamma-aminobutyric acid (Gaba) is a non-proteinogenic amino acid that is widely present in microorganisms, plants, and vertebrates. So far, Gaba is well known as a main inhibitory neurotransmitter in the central nervous system. Its physiological roles are related to the modulation of synaptic transmission, the promotion of neuronal development and relaxation, and the prevention of sleeplessness and depression. Besides, various pharmaceutical properties of Gaba on non-neuronal peripheral tissues and organs were also reported due to anti-hypertension, anti-diabetes, anti-cancer, antioxidant, anti-inflammation, anti-microbial, anti-allergy, hepato-protection, reno-protection, and intestinal protection. Therefore, Gaba may be considered as potential alternative therapeutics for prevention and treatment of various diseases. Accordingly, this updated review was mainly focused to describe the pharmaceutical properties of Gaba as well as emphasize its important role regarding human health.

Keywords: anti-hypertension; bioactivity; Gaba; Gaba-rich product; health benefit

---

1. Introduction

Gamma-aminobutyric acid (Gaba) is a non-protein amino acid that is widely distributed in nature. Especially, Gaba is present in high concentrations in different brain regions [1]. Besides, it was also found in various foods such as green tea, soybean, germinated brown rice, kimchi, cabbage pickles, yogurt, etc. Generally, Gaba was produced by L-glutamic acid under the catalyzation of glutamic acid decarboxylase [2]. In the nervous system, newly synthesized Gaba is packaged into synaptic vesicles and then released into the synaptic cleft to diffuse to the target receptors on the postsynaptic surface [3]. Numerous studies have identified two distinct classes of Gaba receptor including $Gaba_A$ and $Gaba_B$ [4]. These receptors are different due to their pharmacological, electrophysiological, and biochemical properties. $Gaba_A$ receptor is Gaba-gated chloride channels located on the postsynaptic membrane, while $Gaba_B$ receptor is G protein-coupled receptors located both pre- and postsynaptic.

Gaba is well known as the major inhibitory neurotransmitter in the mammalian central nervous system. It was reported to play vital roles in modulating synaptic transmission, promoting neuronal development and relaxation, and preventing sleeplessness and depression [5–9]. Notably, various biological activities of Gaba were documented due to anti-hypertension, anti-diabetes, anti-cancer, antioxidant, anti-inflammation, anti-microbial, and anti-allergy. Moreover, Gaba was also reported as a protective agent of liver, kidney, and intestine against toxin-induced damages [10]. In this contribution, the pharmaceutical properties of Gaba on non-neuronal peripheral tissues and organs were mainly focused to emphasize its beneficial role in prevention and treatment of various diseases.

2. Pharmaceutical Properties of Gaba

2.1. Neuroprotective Effect

It has been reported that the damage of nervous tissue triggers inflammatory response, causing the release of various inflammatory mediators such as reactive oxygen species (ROS), nitric oxide, and cytokines. These mediators can cause several neuronal degenerations in the central nervous system such as Alzheimer's, Parkinson's, and multiple sclerosis [11,12]. So far, numerous studies have been reported regarding the important roles of Gaba on neuro-protection against the degeneration induced by toxin or injury (Figure 1 and Table 1). According to Cho et al. (2007), Gaba produced by the kimchi-derived *Lactobacillus buchneri* exhibited a protective effect against neurotoxic-induced cell death [13]. Moreover, Gaba-enriched chickpea milk can protect neuroendocrine PC-12 cells from $MnCl_2$-induced injury, improve cell viability, and reduce lactate dehydrogenase release [14]. On the other hand, Zhou and colleagues have determined that Gaba receptor agonists also possessed neuroprotective effect against brain ischemic injury. Both $Gaba_A$ and $Gaba_B$ receptor agonist (muscimol and baclofen) could significantly protect neurons from the death induced by ischemia through increasing nNOS (Ser847) phosphorylation [15]. Likewise, the administration of $Gaba_B$ receptor agonist baclofen significantly alleviated neuronal damage and suppressed cytodestructive autophagy via up-regulating the ratio of Bcl-2/Bax and increasing the activation of Akt, GSK-3β, and ERK [16]. Additionally, co-activation of Gaba receptor agonists (muscimol and baclofen) resulted in the attenuation of Fas/FasL apoptotic signaling pathway, inhibition of the kainic acid-induced increase of thioredoxin reductase activity, the suppression of procaspase-3 activation, and the decrease in caspase-3 cleavage. It indicates that co-activation of Gaba receptor agonists results in neuroprotection by preventing caspase-3 denitrosylation in kainic acid-induced seizure of rats [17].

Table 1. Neuroprotective effect of Gaba.

STT	Source	Dose, Model	Time of Treatment/ Administration	Effect	Ref.
1	Kimchi-derived *Lactobacillus buchneri*	100 µg/mL, neuronal cells	24 h	Preventing neurotoxic-induced cell death	[13]
2	*Lactobacillus plantarum*-fermented chickpea milk	537.23 mg/L, PC12 cells	30 min	Preventing $MnCl_2$-induced injury	[14]
3	Gaba receptor agonist	Muscimol (1 mg/kg) and baclofen (20 mg/kg), rat	30 min	Preventing brain ischemic injury and decreasing apoptosis	[15,17]
4	Gaba receptor agonist	Baclofen (10 mL/kg), rat	Once daily/five weeks	Alleviating neuronal damage and suppressing cytodestructive autophagy	[16]

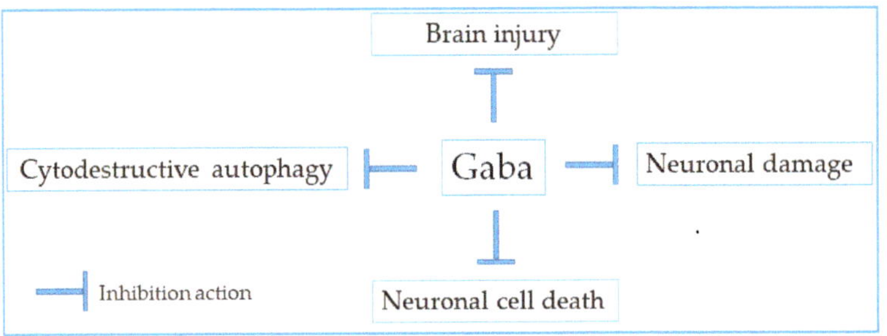

Figure 1. Therapeutic targets for neuroprotective activity of Gaba.

2.2. Neurological Disorder Prevention

Neurologic disorder is associated to dysfunction in part of the brain or nervous system, resulting in physical or psychological symptoms. It includes epilepsy, Alzheimer's disease, cerebrovascular diseases, multiple sclerosis, Parkinson's disease, neuroinfections, and insomnia [18]. It was evidenced that Gaba can suppress neurodegeneration and improve memory as well as cognitive functions of the brain (Figure 2 and Table 2). According to Okada et al. (2000), the usefulness of Gaba-enriched rice germ on sleeplessness, depression, and autonomic disorder was examined [19]. Twenty female patients were administered by Gaba-rich rice germ for three times per day. It was observed that the most common mental symptoms during the menopausal and pre-senile period such as sleeplessness, somnipathy, and depression were remarkably improved in more than 65% of the patients with such symptoms. Likewise, oral administration of Gaba-rich Monascus-fermented product exhibited the protective effect against depression in the forced swimming rat model. Its antidepressant effect was suggested due to recovering the level of monoamines norepinephrine, dopamine, and 5-hydroxytryptamine in the hippocampus [20]. Meanwhile, Yamatsu et al. (2016) reported that Gaba administration significantly shortened sleep latency and increased the total non-rapid eye movement sleep time, indicating the essential role of Gaba in the prevention of a sleep disorder [21]. Moreover, the mixture of Gaba and ʟ-theanine could decrease sleep latency, increase sleep duration, and up-regulate the expression of Gaba and glutamate GluN1 receptor subunit [22]. On the other hand, the electroencephalogram assay has revealed the significantly roles of Gaba in increasing alpha waves, decreasing beta waves, and enhancing IgA levels under stressful conditions. It indicates that Gaba is able to induce relaxation, diminish anxiety, and enhance immunity under stressful conditions [23]. The administration of Gaba-enriched product fermented by kimchi-derived lactic acid bacteria also improved long-term memory loss recovery in the cognitive function-decreased mice and increased the proliferation of neuroendocrine PC-12 cells in vitro [24]. Moreover, the Gaba-enriched fermented *Laminaria japonica* (GFL) provided a protective effect against cognitive impairment associated with dementia in the elderly [25]. In addition, Reid and colleagues have shown that GFL could improve cognitive impairment and neuroplasticity in scopolamine- and ethanol-induced dementia model mice [26]. Especially, GFL was effective in increasing serum brain-derived neurotrophic factor level that associated with lower risk for dementia and Alzheimer's disease in middle-aged women [27]. These results indicate that the use of Gaba-enriched functional foods may improve depression, sleeplessness, cognitive impairment, and memory loss.

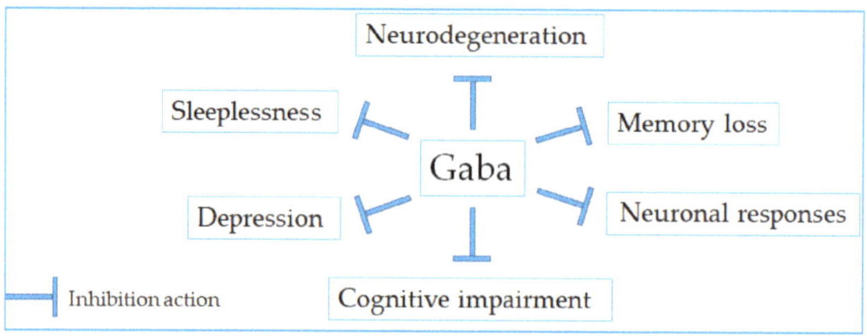

Figure 2. Preventive action of Gaba on neurological disorders.

Table 2. Neurological disorder prevention of Gaba.

STT	Source	Dose/Model	Time of Treatment/ Administration	Effect	Ref.
1	Gaba-enriched rice germ	26.4 mg/3 times/day, patient	N/A	Improving sleeplessness, somnipathy, and depression	[19]
2	Gaba-rich Monascus-fermented product	2.6 mg/kg, rat	30 days	Preventing depression	[20]
3	Gaba powder from natural fermentation using lactic acid bacteria	100 mg Gaba/day, Japanese volunteers	1 week	Prevention of sleep disorder	[21]
4	Gaba (90.8%) and l-theanine (99.3%) was supplied by Neo Cremar Co. Ltd. (Seoul, Korea) and BTC Co. Ltd. (Ansan, Korea), respectively	Gaba/L-theanine mixture (100/20 mg/kg)/day, mice and rat	9 days	Decreasing sleep latency and increasing sleep duration	[22]
5	Gaba from natural fermentation using lactic acid bacteria (Pharma-GABA, Pharma Foods International Co., Japan)	Gaba/L-theanine mixture (100/200 mg/kg)/day Japanese volunteers	7 days	Increasing relaxation, diminishing anxiety, and enhancing immunity	[23]
6	Gaba-enriched product fermented by kimchi-derived lactic acid bacteria	46.69 mg/mL Gaba, mice and PC-12 cells	24 h	Improving long-term memory loss and increasing neuronal cell proliferation	[24]
7	Gaba-enriched fermented *Laminaria japonica* product	1.5 g/day, volunteers	6 weeks	Preventing cognitive impairment in the elderly	[25]

2.3. Anti-Hypertensive Effect

Hypertension is known to relate to a high blood pressure condition, causing various cardiovascular diseases such as ischemic and hemorrhagic stroke, myocardial infarction, and heart and kidney failure [28]. Particularly, angiotensin-I converting enzyme (ACE) was revealed to play an important role in the regulation of blood pressure via converting angiotensin I into the potent vasoconstrictor angiotensin II [29]. Hence, ACE is one of the among therapeutic targets for the control of hypertension. According to Nejati et al. [30], the milk fermented by *Lactococcus lactis* DIBCA2 and *Lactobacillus plantarum* PU11 exhibited an ACE inhibitory activity up to an IC_{50} value of 0.70 ± 0.07 mg/mL. Similarly, high ACE inhibitory activity was also observed by Gaba, which was achieved from *L. plantarum* NTU 102-fermented milk [31]. Moreover, *L. brevis*-fermented soybean containing approximately 1.9 g/kg Gaba was found to possess higher ACE inhibitory activity than the traditional soybean product [32]. Besides, the fermentation of a soybean solution by kimchi-derived lactic acid bacteria in the optimized condition has achieved a Gaba content of up to 1.3 mg/g soybean seeds, and its ACE inhibitory activity was observed up to 43% as compared to the control [33]. Notably, high Gaba content (10.42 mg/g extract) and significant ACE inhibitory activity (92% inhibition) was also determined by the fermented lentils [34].

On the other hand, the anti-hypertensive activity of Gaba was also reported in numerous studies using different experimental models (Table 3). Kimura et al. [35] have investigated the effect of Gaba on blood pressure in spontaneously hypertensive rats. It was observed that the intraduodenal administration of Gaba (0.3 to 300 mg/kg) caused a dose-related decrease in the blood pressure in 30 to 50 min. The hypotensive effect of Gaba was suggested due to attenuating a sympathetic transmission through the activation of the $Gaba_B$ receptor at presynaptic or ganglionic sites. Moreover, the lowering effect of Gaba-enriched dairy product on the blood pressure of spontaneously hypertensive and normotensive Wistar-Kyoto rats was also determined [36]. Notably, the clinical trial has confirmed that daily supplementation of 80 mg of Gaba was effective in the reduction of blood pressure in adults with mild hypertension [37]. Therefore, the consumption of Gaba-enriched dairy product would be beneficial for the down-regulation of hypertension. Indeed, the administration of Gaba-enriched rice grains brings about 20 mmHg decrease in blood pressure in spontaneously hypertensive rats, while there was no significant hypotensive effect in normotensive rats [38]. Likewise, the significant anti-hypertensive activity and the serum cholesterol-lowering effect of Gaba-rich brown rice were shown in spontaneously hypertensive rats as compared to the control [39,40]. In the clinical trial, the effects of Gaba-enriched white rice on blood pressure in 39 mildly hypertensive adults has been examined in a randomized, double blind, placebo-controlled study [41]. It was revealed that the consumption of the Gaba rice could improve the morning blood pressure as compared with the placebo rice after the 1st week and during the 6th and 8th weeks. In the same trend, Tsai and colleagues have determined that Gaba-enriched Chingshey purple sweet potato-fermented milk by lactic acid bacteria (*L. acidophilus* BCRC 14065, *L. delbrueckii* ssp. lactis BCRC 12256, and *L. gasseri* BCRC 14619) was able to reduce both systolic blood pressure and diastolic blood pressure in spontaneously hypertensive rats [42]. The alleviative effect of probiotic-fermented purple sweet potato yogurt on cardiac hypertrophy in spontaneously hypertensive rat hearts was also further determined by Lin and colleagues [43].

In addition, the other Gaba-rich products from bean, tomato, and bread were also reported to be effective in the attenuation of hypertension in vivo. Definite decreases in systolic and diastolic blood pressure values and blood urea nitrogen level were achieved in spontaneously hypertensive rats fed with Gaba-enriched beans [44,45]. Likewise, the anti-hypertensive activity of a Gaba-rich tomato was evidenced to decrease blood pressure in spontaneously hypertensive rats significantly [46]. Moreover, the blood pressure of patients with pre- or mild- to moderate hypertension was significantly decreased during the consumption of 120 g/day of Gaba-rich bread [47]. Accordingly, Gaba-enriched dairy foods may be preferred to use for anti-hypertensive therapeutics.

Table 3. Anti-hypertensive effect of Gaba.

STT	Source	Dose/Model	Time of Treatment/ Administration	Effect	Ref.
1	Milk fermented by *Lactococcus lactis* DIBCA2 and *Lactobacillus plantarum* PU11	0.70 mg/ml	5 min	Inhibiting 50% ACE activity	[30]
2	Gaba form LAB-fermented soybean	1.3 mg Gaba/g soybean	10 min	Inhibiting 43% ACE activity	[33]
3	Gaba from the fermented lentils	10.42 mg Gaba/g extract	60 min	Inhibiting 92% ACE activity	[34]
4	Gaba from Wako Pure Chemicals (Tokyo)	0.3 to 300 mg Gaba/kg, rat	Every 20 min for i.v. administration	Decreasing blood pressure	[35]
5	Gaba from skim cows' milk fermented with *Lactobacillus casei* strain Shirota and *Lactococcus lactis* YIT 2027	5 mL (102 mg Gaba/kg) of the fermented solution/kg body weight, rat	10 h	Lowering blood pressure	[36]
6	Gaba-enriched rice grains	0.1 mg–0.5 mg Gaba/kg, rat	6 weeks	Decreasing blood pressure	[38]
7	Gaba-enriched white rice	150 g of Gaba-enriched white rice (11.2 mg Gaba/100 g rice), volunteers	8 weeks	Decreasing blood pressure	[41]
8	Gaba-enriched Chingshey purple sweet potato-fermented milk by lactic acid bacteria	2.5-mL dose of fermented-milk, rat	8 weeks	Reducing both systolic blood pressure and diastolic blood pressure	[42]
9	Gaba from probiotic-fermented purple sweet potato yogurt	1500 µg/2.5 mL/kg, rat	8 weeks	Alleviating cardiac hypertrophy	[43]
10	Gaba-rich tomato	2–10 g/kg, rat	2–24 h	Decreasing blood pressure	[46]
11	Gaba-rich bread	120 g/day, patient	3 days	Decreasing blood pressure	[47]

2.4. Anti-Diabetic Effect

Diabetes is an endocrine disorder that is associated with dysregulation of carbohydrate metabolism and deficiency of insulin secretion or insulin action, causing chronic hyperglycemia [48]. So far, diabetic diseases can be managed by pharmacologic interventions [49]. However, the lowering blood glucose effect of pharmacological drugs is accompanied with various disadvantages such as drug resistance, side effects, and even toxicity [50]. Therefore, the proper diet and exercise have been recommended and preferred as alternative therapeutics for the regulation of diabetic diseases. Notably, Gaba and Gaba-enriched natural products have been evidenced as effective agents in lowering blood glucose, attenuating insulin resistance, stimulating insulin release, and preventing pancreatic damage (Figure 3 and Table 4). Soltani and colleagues have shown that Gaba enhanced islet cell function via producing membrane depolarization and Ca$^{(2+)}$ influx, activating PI3-K/Akt-dependent growth and survival pathways, and restoring the β-cell mass [51]. Moreover, Gaba preferentially up-regulated pathways linked to β-cell proliferation and rose to a distinct subpopulation of β cells with a unique transcriptional signature, including urocortin3, wnt4, and hepacam2 [52]. Especially, the combined use of Gaba and sitagliptin was superior in increasing β-cell proliferation, reducing cell apoptosis, and suppressing α-cell mass [53]. On the other hand, Gaba was found to enhance insulin secretion in pancreatic INS-1 β-cells [54]. In the pre-clinical trial model, Gaba administration could decrease the ambient blood glucose level and improve the glucose excursion rate in streptozotocin-induced diabetic mice [53]. Furthermore, oral treatment with Gaba significantly reduced the concentrations of fasting blood glucose, improved glucose tolerance and insulin sensitivity, and inhibited the body weight gain in the high fat diet-fed mice [55]. Notably, Gaba potentially inhibited the diabetic complication related to the nervous system via suppressing the Fas-dependent and mitochondrial-dependent apoptotic pathway in the cerebral cortex [56].

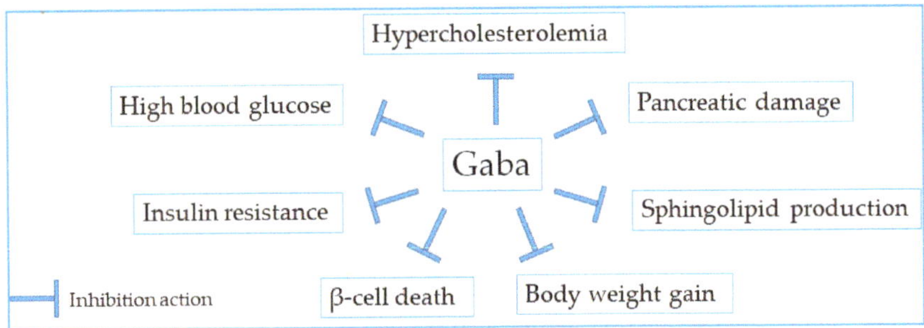

Figure 3. Therapeutic targets for anti-diabetic activity of Gaba.

The fact that the germination of rice and the fermentation of foods are accompanied with the increase in Gaba content [57,58], therefore, the pre- and germinated rice and fermented foods were highly appreciated for their roles in positive regulation of diabetes and its complication. According to Hagiwara and colleagues, the feeding of pre-germinated brown rice diet to diabetic rats significantly decreased blood glucose, adipocytokine PAI-1 concentration, and plasma lipid peroxide [59]. Moreover, pre-germinated brown rice lowered HbA(1c) and adipocytokine (TNF-α and PAI-1) concentration and increased the adiponectin level in type-2 diabetic rats, leading to the prevention of potential diabetic complications [60]. In addition, high fat diet-induced diabetic pregnant rats fed with the germinated brown rice lead to the increase in adiponectin levels and the reduction of insulin, homeostasis model assessment of insulin resistance, leptin, and oxidative stress in their offspring [61]. On the other hand, blackish purple pigmented rice with a giant embryo significantly decreased blood glucose and plasma insulin levels, adipokine concentrations, and hepatic glucose-regulating enzyme activities in ovariectomized rats [62]. Meanwhile, glucose homeostasis was greatly improved through the intervention of Gaba-enriched wheat bran in the context of a high-fat diet rat [63]. The supplement of Gaba-enriched rice bran to obese rats also exhibited an efficient effect on lowering serum sphingolipids, a marker of insulin resistance [64]. In clinical trials, Ito and colleagues have suggested that the intake of pre-germinated brown rice was effective in lowering postprandial blood glucose concentration without insulin secretion increase [65]. Likewise, Hsu et al. [66] and Suzuki et al. [67] have confirmed that pre-germinated brown rice decreased blood glucose and hypercholesterolemia in type 2 diabetes patients.

Beside germinated rice, fermented foods are also known to contain a significant amount of Gaba and possess potential anti-diabetic activity. The oral administration of hot water extract of the fermented tea obtained by tea-rolling processing of loquat (*Eriobotrya japonica*) significantly decreased the blood glucose level and serum insulin secretion in maltose-loaded Sprague–Dawley rats [68]. Similarly, anti-diabetic effects of green tea fermented by cheonggukjang was observed via decreasing water intake and lowering blood glucose and HbA1c levels in diabetic mice [69]. In addition, mung bean fermented by *Rhizopus* sp. [70], yogurt fermented by *Streptococcus salivarius* subsp. thermophiles fmb5 [71], and soybean extract fermented by *Bacillus subtilis* MORI [72] could enhance their anti-hyperglycemic effect via reducing blood glucose, HbA1c, cholesterol, triglyceride, and low–density lipoprotein levels in diabetic mice. In the same trend, the milk fermented by commercial strain YF-L812 (*S. thermophilus, L. delbrueckii* subsp. *bulgaricus*), standard strains. *B. breve* KCTC 3419, and *L. sakei* LJ011. Fermented milk was effective in decreasing fasting blood glucose, serum insulin, leptin, glucose and insulin tolerance, total cholesterol, triglycerides, and low density lipoprotein cholesterol [73]. Especially, the consumption of probiotic-fermented milk (kefir) by type 2 diabetic patents lowered HbA1C level, homeostatic model assessment of insulin resistance, and homocysteine amount [74,75]. Accordingly, the germinated rice and fermented foods, which contain a high amount of Gaba, could be used as anti-diabetic functional food for maintaining health and preventing complications in type 2 diabetes.

Table 4. Anti-diabetic effect of Gaba.

STT	Source	Dose/Model	Time of Treatment/ Administration	Effect	Ref.
1	Gaba (Source: N/A)	Dose: N/A, mice	8–15 weeks	Activating PI3-K/Akt-dependent growth and survival pathways and restoring the β-cell mass	[51]
2	Gaba (MilliporeSigma, Burlington, MA, USA)	Gaba (6 mg/mL/day), mice	10 weeks	Up-regulating β-cell proliferation and rising a distinct subpopulation of β cells	[52]
3	Gaba (Sigma, St. Louis, USA)	Gaba (2 mg/mL/day), mice	20 weeks	Reducing the concentrations of fasting blood glucose, improving glucose tolerance and insulin sensitivity, and inhibiting the body weight gain	[55]
4	Gaba from pre-germinated brown rice	Pre-germinated brown rice (1387–1546 g/day), rat	7 weeks	Decreasing blood glucose, adipocytokine PAI-1 concentration, and plasma lipid peroxide	[59]
5	Gaba from germinated brown rice	Gaba (200 mg/kg/day), rat offspring	8 weeks	Increasing adiponectin levels and reducing insulin resistance and oxidative stress	[61]
6	Gaba from blackish purple pigmented rice with a giant embryo	Diet supplemented with either 20% (*w/w*) germinated Keunnunjami rice powder, rat	8 weeks	Decreasing blood glucose and plasma insulin levels, adipokine concentrations, and hepatic glucose-regulating enzyme activities	[62]
7	Gaba-enriched wheat bran	15% Gaba-enriched bran, rat	8 weeks	Improving glucose homeostasis	[63]

Table 4. *Cont.*

STT	Source	Dose/Model	Time of Treatment/ Administration	Effect	Ref.
8	Gaba from pre-germinated brown rice	The test sample contained 50 g of available carbohydrate per day for each volunteer (185 g of pre-germinated brown rice), volunteers	7 weeks	Lowering postprandial blood glucose concentration without insulin secretion increase	[65]
9	Gaba from pre-germinated brown rice	180 g of the cooked rice/three times per day, patient	14 weeks	Decreasing blood glucose and hypercholesterolemia	[66]
10	Fermented tea product	50 mg/kg, rat	120 min	Decreasing blood glucose level	[68]
11	Mung bean fermented by *Rhizopus* sp.	200 mg/kg and 1000 mg/kg, mice	240 min	Reducing blood glucose, HbA1c, cholesterol, triglyceride, and low-density lipoprotein levels	[70]
12	Yogurt fermented by *Streptococcus salivarius* subsp.	Gaba orally at a dose of 2 g/L or 4 g/L	6 weeks	Reducing blood glucose, HbA1c, cholesterol, triglyceride, and low-density lipoprotein levels	[71]
13	Soybean extract fermented by *Bacillus subtilis* MORI	500 mg/kg, mice	8 weeks	Reducing blood glucose, HbA1c, cholesterol, triglyceride, and low-density lipoprotein levels	[72]
14	Milk fermented by strain YF-L812 (*S. thermophilus, L. delbrueckii* subsp. *bulgaricus*), standard strains. *B. breve* KCTC 3419, and *L. sakei* LJ011. FM	Fermented milk 0.2% and 0.6%/kg/day, mice	6 weeks	Decreasing fasting blood glucose, serum insulin, insulin tolerance, total cholesterol, triglycerides, and LDL cholesterol	[73]

2.5. Anti-Cancer Effect

Cancer is involved in the unregulated cell proliferation, apoptosis suppression, invasion, and metastasis [76]. Current cancer therapies are related to surgery, radiation treatment, and chemotherapy treatment, which are widely applied for treatment of all kinds of cancers. However, these therapies possess major disadvantages including cancer recurrence, drug resistance, and side effects. Hence, the discovery of alternative medicines with desirable properties is always necessary. In this regard, Gaba was emerged as a promising compound that is able to regulate cancer due to the induction of apoptosis and inhibition of proliferation and metastasis (Table 5). Gaba-enriched brown rice extract significantly retarded the proliferation rates of L1210 and Molt4 leukemia cells and enhanced apoptosis of the cultured L1210 cells [77]. Moreover, Schuller et al. [78] suggested that Gaba had a tumor suppressor function in small airway epithelia and pulmonary adenocarcinoma, providing the approach for the prevention of pulmonary adenocarcinoma in smokers. According to Huang and colleagues, Gaba was determined to inhibit the activity and expression of MMP-2 and MMP-9 in cholangiocarcinoma QBC939 cells, suggesting its role in prevention of invasion and metastasis in cancer [79]. Song and colleagues also found the inhibitory effects of Gaba on the proliferation and metastasis of colon cancer cells (SW480 and SW620 cells) due to the up-pressing cell cycle progression (G2/M or G1/S phase), attenuating mRNA expression of EGR1-NR4A1 and EGR1-Fos axis, and disrupting MEK-EGR1 signaling pathway [80]. Especially, the co-treatment of Gaba and Celecoxib significantly inhibited systemic and tumor VEGF, PGE$_2$, and cAMP molecules and down-regulated COX-2 and p-5-LOX protein in pancreatic cancer cells [81]. Moreover, the prolonged administration of Gaba at 1000 mg/kg body weight significantly decreased the number of gastric cancers of the glandular stomach in Wk 52 rats. In parallel, the histological method also revealed the role of Gaba on decreasing the labeling index of the antral mucosa and increasing the serum gastrin level [82]. Likewise, the pre-treatment of Gaba also significantly reduced intrahepatic liver metastasis and primary tumor formation in mice and inhibited human liver cancer cell migration and invasion via the induction of liver cancer cell cytoskeletal reorganization [83]. Meanwhile, the increase in the activity of Gaba$_A$ receptor contributed to the down-regulation of alpha-fetoprotein mRNA expression and cell proliferation in malignant hepatocyte cell line [84].

Table 5. Anti-cancer effect of Gaba.

STT	Source	Dose/Model	Time of Treatment/ Administration	Effect	Ref.
1	Gaba-enriched brown rice extract	20 µL extract/well (1 × 10^5 cells/200 mL/well), leukemia cells and HeLa cells	48 h	Retarding the proliferation rates of leukemia cells and enhancing apoptosis of leukemia cells	[77]
2	Gaba from Sigma Company (St. Louis, MO, USA)	Gaba (1–1000 µmol/L), cholangiocarcinoma QBC939 cells	24 h	Inhibiting the activity and expression of MMP-2 and MMP-9	[79]
3	Gaba was purchased from Sigma-Aldrich, Shanghai, China	Gaba (100 µM), Colon cancer cells	72 h	Inhibiting on cell proliferation and metastasis	[80]
4	Gaba from Sigma Company (St. Louis, MO, USA)	Gaba (1000 mg/kg), rat	25 weeks	Decreasing the number of gastric cancers of the glandular stomach	[82]
5	Gaba from Sigma Company (St. Louis, MO, USA)	Gaba (10 µM), Human liver cancer cells	24 h	Reducing intrahepatic liver metastasis and inhibiting human liver cancer cell migration and invasion	[83]

2.6. Antioxidant Effect

The free radicals contain one or more unpaired electrons that are generated from the living organisms and external sources. The high level of free radicals could cause the damage of the body's tissues and cells, leading to human aging and various diseases [85,86]. Thus, consumption of natural products with high anti-oxidant effect is useful for the prevention of free radical-caused diseases [86]. Herein, the antioxidant property of Gaba has been evidenced in numerous studies (Figure 4). It was shown that Gaba was able to trap the reactive intermediates during lipid peroxidation and react readily with malondialdehyde under physiological conditions [87]. Moreover, the administration of Gaba significantly decreased malondialdehyde concentration and increased the activity of superoxide dismutase and glutathione peroxidase in the cerebral cortex and hippocampus of acute epileptic state rats [88]. In other studies, the protective effect of Gaba against H$_2$O$_2$-induced oxidative stress in pancreatic cells [89] and human umbilical vein endothelial cells [90] was observed via reducing cell death, inhibiting reactive oxygen species (ROS) production, and enhancing antioxidant defense systems. Similarly, gamma rays-induced oxidative stress in the small intestine of rats was significantly ameliorated via decreasing malondialdehyde and advanced oxidation protein productions, increasing catalase and glutathione peroxidase activities, preventing mucosal damage and hemorrhage, and inducing the regeneration of the small intestinal cells [91]. Gaba also attenuated brain oxidative damage associated with insulin alteration in streptozotocin-treated rats [92]. On the other hand, Gaba from *L. brevis*-fermented sea tangle solution was observed to exhibit stronger antioxidant activity than positive control BHA in scavenging DPPH and superoxide radicals and inhibiting xanthine oxidase [93]. Meanwhile, the Gaba-rich germinated brown rice extract considerably scavenged hydroxyl radical and thiobarbituric acid-reactive substances in both cell-free medium and post-treatment culture media, indicating its radical scavenging capacity in both direct and indirect action [94]. Recently, brew-germinated pigmented rice vinegar was also suggested as a new product with high antioxidant activity [95].

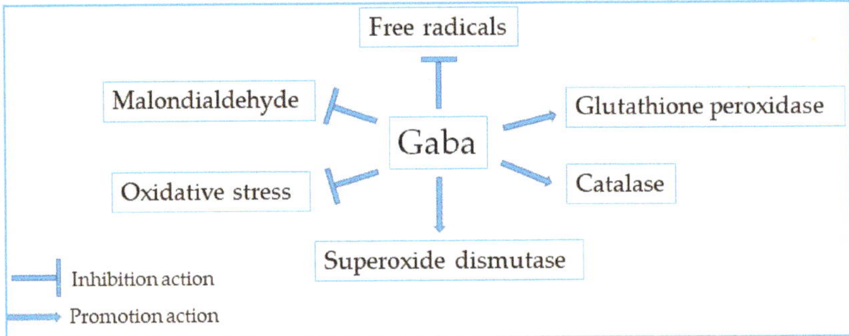

Figure 4. Modulatory activity of Gaba for antioxidant promotion.

2.7. Anti-Inflammatory Effect

Inflammation response is triggered by the stimulation of various factors such as physical damage, ultra violet irradiation, microbial invasion, and immune reactions [96]. It is associated with the production of a large range of pro-inflammatory mediators such cytokine, NO, and PGE$_2$ [97]. Notably, Gaba was indicated as an inhibitor of inflammation via decreasing pro-inflammatory mediator production and ameliorating inflammatory symptom (Figure 5). At the early time, Han et al. [98] have determined the anti-inflammatory activity of Gaba via inhibiting the production and expression of iNOS, IL-1β, and TNF-α in LPS-stimulated RAW 264.7 cells. As the result, it contributed to the reduction of the whole healing period and enhancement of wound healing at the early stage. Likewise, Gaba suppressed inflammatory cytokine production and NF-kB inhibition in both lymphocytes and pancreatic islet beta cells [99]. Recently, Gaba-enriched sea tangle *L. japonica*, Gaba-rich germinated brown rice, and Gaba-rich red microalgae *Rhodosorus marinus* were reported for their inhibitory capacities on inflammatory response. Gaba-enriched sea tangle *L. japonica* extract suppressed nitric oxide production and inducible nitric oxide synthase expression in LPS-induced mouse macrophage RAW 264.7 cells [100]. Gab-rich germinated brown rice inhibited IL-8 and MCP-1 secretion and ROS production from Caco-2 human intestinal cells activated by H$_2$O$_2$ and IL-1β [101]. Gaba-rich red microalgae *Rhodosorus marinus* extract negatively modulated expression and release of pro-inflammatory IL-1α in phorbol myristate acetate-stimulated normal human keratinocytes, therefore indicating the potential treatment of sensitive skins, atopia, and dermatitis [102]. Besides, the roles of Gaba in the attenuation of gut inflammation and improvement of gut epithelial barrier were suggested via inhibiting IL-8 production and stimulating the expression of tight junction proteins as well as the expression of TGF-β cytokine in Caco-2 cells [103].

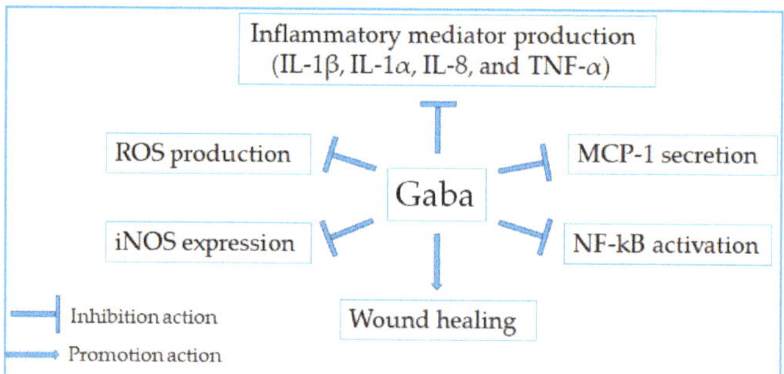

Figure 5. Therapeutic targets for anti-inflammatory activity of Gaba.

2.8. Anti-Microbial Effect

Gaba tea is a kind of Gaba-enriched tea by the repeating treatments of alternative anaerobic and aerobic conditions. The Gaba tea extract exhibited inhibitory activity against *Vibrio parahaemolyticus*, *Staphylococcus aureus*, *Bacillus cereus*, *Salmonella typhimurium*, and *Escherichia coli* [104]. Gaba could increase *Pseudomonas aeruginosa* virulence due to stimulation of cyanogenesis, reduction in oxygen accessibility, and overexpression of oxygen-scavenging proteins. Gaba also promotes specific changes in the expression of thermostable and unstable elongation factors involved in the interaction of the bacterium with the host proteins [105]. Recently, the role of Gaba in anti-microbial host defenses was elucidated by Kim and colleagues [106]. Treatment of macrophages with Gaba enhanced phagosomal maturation and anti-microbial responses against mycobacterial infection. This study identified the role of Gabaergic signaling in linking anti-bacterial autophagy to enhance host innate defense against intracellular bacterial infection including *Mycobacteria*, *Salmonella*, and *Listeria*.

2.9. Anti-Allergic Effect

Allergy is a disorder of the immune system associating with an exaggerated reaction of the immune system to harmless environmental substances. Allergic reaction is characterized by the excessive activation of mast cells and basophils, leading to release various mediators such as histamine and an array of cytokines [107]. Among them, histamine is considered as the major target for potential anti-allergic therapeutics. Herein, the inhibitory activity of Gaba on histamine release from the activated mast cells was investigated in vitro [108,109]. Rat basophilic leukemia cells and rat peritoneal exudate cells sensitized with anti-dinitrophenyl (DNP) IgE and challenged with DNP-conjugated bovine serum albumin resulted in the release of histamine in a cell culture medium. However, IgE-mediated histamine release was inhibited by Gaba treatment in both cells. Conversely, the inhibitory activities of Gaba were lowered by the addition of CGP35348, a Gaba$_B$ receptor antagonist. It indicated that Gaba inhibited degranulation from basophils and mast cells via Gaba$_B$ receptor on the cell surface. On the other hand, Hokazono et al. [110] have evaluated the protective effect of Gaba against the development of atopic dermatitis (AD)-like skin lesions in NC/Nga mice. It was observed that Gaba could prevent the development of AD-like skin lesions in mice via alleviating serum immunoglobulin E (IgE) and splenocyte IL-4 production. The combined administration of Gaba and the fermented barley extract remarkably increased splenic cell interferon-γ production, indicating the domination of Th1/Th2 balance to Th1 response. Hence, the simultaneous intake of Gaba and the fermented barley extract was encouraged to ameliorate allergic symptoms such as atopic dermatitis (Figure 6).

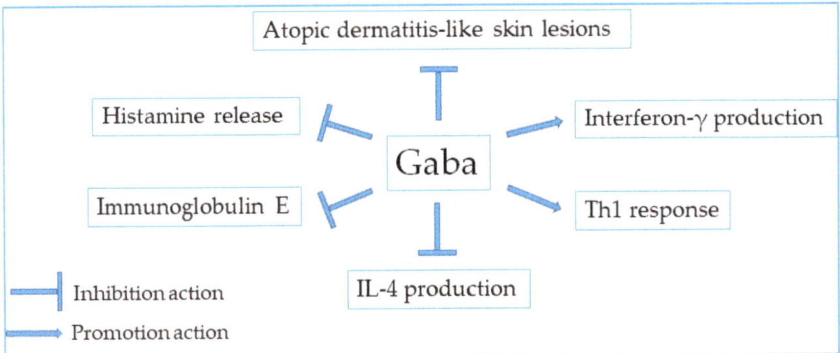

Figure 6. Therapeutic targets for anti-allergic activity of Gaba.

2.10. Hepatoprotective Effect

The long-term use of ethanol can cause liver damage and unfavorable lipid profiles in humans. The toxic acetaldehyde is formed from alcohol under catalysis of alcohol dehydrogenase, causing various adverse effects such as thirst, vomiting, fatigue, headache, and abdominal pain [111]. For the first time, Oh and colleagues have evaluated the protective effect of Gaba-rich germinated brown rice against the toxic consequences of chronic ethanol use [112]. Interestingly, serum low-density lipoprotein cholesterol, liver aspartate aminotransferase, and liver alanine aminotransferase levels were decreased in mice fed both ethanol and brown rice extract for 30 days. Furthermore, the brown rice extract significantly increased serum and liver high-density lipoprotein cholesterol concentrations and reduced liver triglyceride and total cholesterol concentrations. In the same trend, Lee et al. [113] have reported that Gaba-rich fermented sea tangle (GFST) could prevent ethanol and carbon tetrachloride-induced hepatotoxicity in rats. The oral administration of GFST decreased the serum levels of glutamic pyruvate transaminase, gamma glutamyl transpeptidase, and malondialdehyde levels and increased antioxidant enzyme such as superoxide dismutase, catalase, and glutathione peroxidase [113]. Moreover, GFST increased the activities and transcript levels of major alcohol-metabolizing enzymes, such as alcohol dehydrogenase and aldehyde dehydrogenase, and reduced blood concentrations of alcohol and acetaldehyde [114]. In an in vitro study, the protective effects of GFST against alcohol hepatotoxicity in ethanol-exposed HepG$_2$ cells were revealed by preventing intracellular glutathione depletion, decreasing gamma-glutamyl transpeptidase activity, and suppressing cytochrome P450 2E1 enzyme expression [115]. These results indicated that Gaba-rich foods might have a pharmaceutical role in the prevention of chronic alcohol-related diseases (Figure 7).

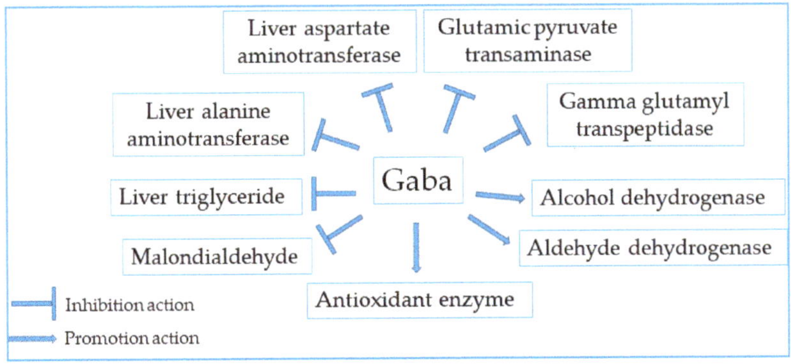

Figure 7. Mechanism of the action of Gaba for hepatoprotection.

2.11. Renoprotective Effect

Acute kidney injury is involved in kidney damage and cell death, causing high morbidity and mortality worldwide [116]. The renoprotective agents derived from natural products may be essential for the prevention or treatment of kidney injury-related diseases. Indeed, numerous studies have evidenced the protective effect of Gaba against acute kidney injury (Figure 8). According to Kim et al. (2004), the physiological changes caused by acute renal failure such as body weight and kidney weight gain, urea nitrogen and creatinine elevation, creatinine clearance reduction, sodium FE(Na) secretion, and urine osmolarity decrease in rats were significantly improved by oral administration of Gaba [117]. Moreover, the status of serum albumin decrease, urinary protein increase, and serum lipid profile was completely improved by Gaba. In addition, Gaba alleviated nephrectomy-induced oxidative stress by increasing superoxide dismutase and catalase, and decreasing lipid peroxidation in rats [118]. Furthermore, Gaba reduced tubular fibrosis, tubular atrophy, and the transforming growth factor-beta1 and fibronectin expression [119]. The acute tubular necrosis was also apparently reduced to normal proximal condition by Gaba treatment [120]. In another study, Talebi and colleagues have shown the protective effect of Gaba on kidney injury induced by renal ischemia-reperfusion in ovariectomized rats via decreasing serum levels of creatinine and blood urea nitrogen, kidney weight, and kidney tissue damage [121]. Meanwhile, the increases in alanine amino transferase and aspartate amino transferase activities, urea and creatinine levels, malondialdehyde and advanced oxidation protein levels, and oxidative damage to the kidney tissues induced by γ-irradiated- and streptozotocin-treated rats were markedly attenuated by Gaba administration in rats [122]. Especially, Gaba was observed to ameliorate kidney injury induced by renal ischemia/reperfusion injury in a gender dependent manner [123]. These results emphasized the protective effect of Gaba against the renal damage involving in renal failure.

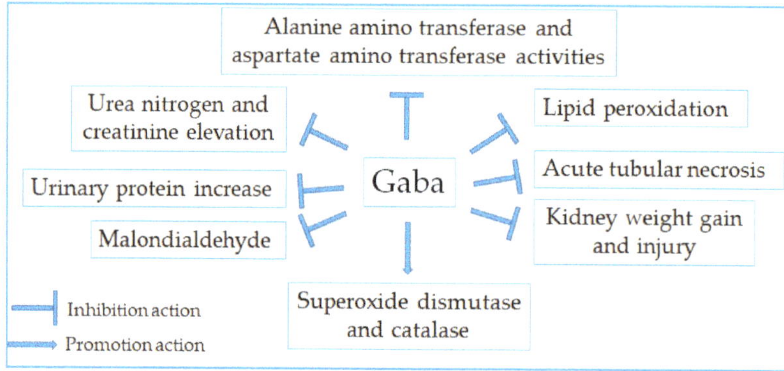

Figure 8. Mechanism of the action of Gaba for renoprotection.

2.12. Intestinal Protective Effect

Chen and colleagues have examined the beneficial roles of Gaba on intestinal mucosa in vivo [124,125]. It was shown that heat stress-induced chicken decreased the activity of Na^+-K^+-ATPase, maltase, sucrase, and alkaline phosphatase enzymes in intestinal mucosa [124]. Moreover, heat stress caused the marked decline in villus length, mucosa thickness, intestinal wall thickness, and crypt depth in the duodenum and ileum [125]. However, the treatment of Gaba administration markedly increased the activity of maltase, sucrase, alkaline phosphatase, and Na^+-K^+-ATPase [124]. Furthermore, Gaba enhanced villus length, mucosa thickness, intestinal wall thickness, and crypt depth in the duodenum and ileum [125]. It indicated that Gaba could effectively alleviate heat stress-induced damages of the intestinal mucosa. In a further study, they investigated the effect of Gaba supplementation on the growth performance, intestinal immunity, and gut microflora of the weaned piglets [126]. Notably, Gaba supplementation improved the growth performance, inhibited proinflammatory cytokines (IL-1 and IL-18) expression, promoted anti-inflammatory cytokines (IFN-γ, IL-4, and IL-10) expression, and increased the dominant microbial populations, the community richness, and diversity of the ileal microbiota. On the other hand, Xie and colleagues also investigated the effect of Gaba on colon health in mice [127]. It was observed that the female Kunming mice administrated with Gaba at doses of 40 mg/kg/d for 14 days could increase the concentrations of acetate, propionate, butyrate, and total short chain fatty acids, and decreased pH value in colonic and cecal contents. Recently, Kubota and colleagues have revealed that Gaba attenuated ischemia reperfusion-induced alterations in intestinal immunity via increasing IgA secretion, alpha-defensin-5 expression, and superoxide dismutase activity in the rat small intestine [128]. Besides, Jiang and colleagues also showed the protective effect of Gaba against intestinal mucosal barrier injury of colitis induced by 2,4,6-trinitrobenzene sulfonic acid and alcohol [129]. These results have evidenced the physiological function of Gaba in improvement and promotion of intestinal health.

2.13. Other Pharmaceutical Properties

Yang et al. [130] have examined the modulatory effects of Gaba on cholesterol-metabolism-associated molecules in human monocyte-derived macrophages (HMDMs). It was found that Gaba was effective in the reduction of cholesterol ester in lipid-laden HMDMs via suppressing the expression of scavenger receptor class A, lectin-like oxidized low-density lipoprotein receptor-1, and CD36, and promoting the expression of ATP-binding cassette transporter 1, ATP-binding cassette sub-family G member 1, and scavenger receptor class B type I. Moreover, the production of TNF-α was decreased and the activation of signaling pathways (p38MAPK and NF-κB) was repressed in the presence of Gaba. The inhibitory effect of Gaba on the formation of human macrophage-derived foam cells suggests its role in the prevention of atherosclerotic lesions.

Yang et al. [131] have investigated whether Gaba ameliorate fluoride-induced a thyroid injury in vivo. The model of hypothyroidism was conducted by exposing NaF (50 mg/kg) to adult male mice for 30 days. Thereafter, thyroid hormone production, oxidative stress, thyroid function-associated genes, and side effects during therapy were measured. Interestingly, Gaba supplementation remarkedly promoted the expression of thyroid thyroglobulin, thyroid peroxidase, and sodium/iodide symporter. Moreover, it improved the thyroid redox state, the expression of thyroid function-associated genes, and liver metabolic protection. These findings indicate that Gaba has a therapeutic potential in hypothyroidism.

In regarding to the growth hormone, the oral administration of Gaba was reported to elevate the resting and post-exercise immunoreactive growth hormone and immunofunctional growth hormone concentrations in humans [132]. Moreover, the administration of Gaba is likely to increase the concentrations of plasma growth hormone and the rate of protein synthesis in the rat brain [133,134]. Recently, the role of Gaba in the enhancement of muscular hypertrophy in men after progressive resistance training was also evaluated by Sakashita and colleagues [135]. They found that the combination of Gaba and whey protein was effective in increasing whole body fat-free mass, thus enhancing exercise-induced muscle hypertrophy.

Indeed, the excessive production of free radicals and oxidants causes oxidative stress that damages cell membranes and other structures such as DNA, lipids, and proteins [136]. Particularly, the damage of cell membranes and lipoproteins by hydroxyl and peroxynitrite radicals causes lipid peroxidation and formation of cytotoxic and mutagenic agents such as malondialdehyde and conjugated diene compounds [137]. Moreover, the free radicals and oxidants can change protein structure and lose enzyme activity. Various mutations may also result from oxidants-induced DNA damages. Therefore, oxidative stress can induce a variety of chronic and degenerative diseases such as cancer, cardiovascular disease, neurological disease, pulmonary disease, rheumatoid arthritis, nephropathy, and ocular disease [138]. In this sense, antioxidants play an important role in the neutralization of free radicals, protection of the cells from toxic effects, and prevention of disease pathogenesis [139]. As a result, the antioxidant activity of Gaba may partly contribute to its biological effects such as anti-hypertension, anti-diabetes, anti-cancer, antioxidant, anti-inflammation, anti-microbial, anti-allergy, hepato-protection, reno-protection, and intestinal protection.

3. Conclusions

The fact that consumers have paid much attention to natural products in order to promote and maintain their health. Simultaneously, various functional foods derived from natural products have been developed along with the tendency of consumers. Herein, Gaba has been evidenced as a powerful bioactive compound with numerous health beneficial effects. Thus, the functional foods produced from Gaba are believed to be able to prevent and/or treat different diseases, especially hypertension, diabetes, and neurological disorders. Whereby, the researches into large-scale production, biotechnological techniques, and high Gaba-producing strains will be remarkedly increased in food industry. However, the further testing and validation due to the safety and efficacy of Gaba consumption are necessary in clinical trials.

Author Contributions: We declare that this review was done by the authors named in this article. The review was conceived and designed by D.-H.N. The data were collected and analyzed by D.-H.N. and T.S.V. The manuscript was written by T.S.V. All authors read and approved the manuscript for publication.

Funding: This research is funded by Vietnam National Foundation for Science and Technology Development (NAFOSTED) under grant number 106.02-2018.304.

Acknowledgments: This review is also supported by Nguyen Tat Thanh University, Ho Chi Minh city, Vietnam and Thu Dau Mot University, Binh Duong province, Vietnam.

Conflicts of Interest: There are no conflicts to declare.

References

1. Olsen, R.W.; Betz, H. GABA and glycine. In *Basic Neurochemistry*, 7th ed.; Siegel, G.J., Albers, R.W., Brady, S.T., Price, D.L., Eds.; Elsevier Academic Press: Boston, MA, USA, 2006.
2. Martin, D.L.; Olsen, R.W. (Eds.) *GABA in the Nervous System: The View at 50 Years*; Lippincott Williams & Wilkins: Philadelphia, PA, USA, 2000.
3. Madsen, K.K.; Clausen, R.P.; Larsson, O.M.; Krogsgaard-Larsen, P.; Schousboe, A.; White, H.S. Synaptic and extrasynaptic GABA transporters as targets for anti-epileptic drugs. *J. Neurochem.* **2009**, *109*, 139–144. [CrossRef] [PubMed]
4. Mann, E.O.; Kohl, M.M.; Paulsen, O. Distinct roles of GABA(A) and GABA(B) receptors in balancing and terminating persistent cortical activity. *J. Neurosci.* **2009**, *29*, 7513–7518. [CrossRef] [PubMed]
5. Stagg, C.J.; Bachtiar, V.; Johansen-Berg, H. The role of GABA in human motor learning. *Curr. Biol.* **2011**, *21*, 480–484. [CrossRef] [PubMed]
6. Nuss, P. Anxiety disorders and GABA neurotransmission: A disturbance of modulation. *Neuropsychiatr. Dis. Treat.* **2015**, *11*, 165–175. [PubMed]
7. Wagner, S.; Castel, M.; Gainer, H.; Yarom, Y. GABA in the mammalian suprachiasmatic nucleus and its role in diurnal rhythmicity. *Nature* **1997**, *387*, 598–603. [CrossRef] [PubMed]
8. Gottesmann, C. GABA mechanisms and sleep. *Neuroscience* **2002**, *111*, 231–239. [CrossRef]
9. Plante, D.T.; Jensen, J.E.; Schoerning, L.; Winkelman, J.W. Reduced γ-aminobutyric acid in occipital and anterior cingulate cortices in primary insomnia: A link to major depressive disorder? *Neuropsychopharmacology* **2012**, *37*, 1548–1557. [CrossRef] [PubMed]
10. Rashmi, D.; Zanan, R.; John, S.; Khandagale, K.; Nadaf, A. γ-aminobutyric acid (GABA): Biosynthesis, role, commercial production, and application. In *Study in Natural Products Chemistry*, 1st ed.; Rahman, A., Ed.; Elsevier: 1000 AE Amsterdam, The Netherlands, 2018; Volume 57, pp. 413–452.
11. Kaminsky, N.; Bihari, O.; Kanner, S.; Barzilai, A. Connecting malfunctioning glial cells and brain degenerative disorders. *Genom. Proteom. Bioinform.* **2016**, *14*, 155–165. [CrossRef] [PubMed]
12. Bagli, E.; Goussia, A.; Moschos, M.M.; Agnantis, N.; Kitsos, G. Natural compounds and neuroprotection: Mechanisms of action and novel delivery systems. *In Vivo* **2016**, *30*, 535–547.

13. Cho, Y.R.; Chang, J.Y.; Chang, H.C. Production of gamma-aminobutyric acid (GABA) by *Lactobacillus buchneri* isolated from kimchi and its neuroprotective effect on neuronal cells. *J. Microbiol. Biotechnol.* **2007**, *17*, 104–109. [PubMed]

14. Li, W.; Wei, M.; Wu, J.; Rui, X.; Dong, M. Novel fermented chickpea milk with enhanced level of γ-aminobutyric acid and neuroprotective effect on PC12 cells. *Peer J* **2016**, *4*, e2292. [CrossRef] [PubMed]

15. Zhou, C.; Li, C.; Yu, H.M.; Zhang, F.; Han, D.; Zhang, G.Y. Neuroprotection of gamma-aminobutyric acid receptor agonists via enhancing neuronal nitric oxide synthase (Ser847) phosphorylation through increased neuronal nitric oxide synthase and PSD95 interaction and inhibited protein phosphatase activity in cerebral ischemia. *J. Neurosci. Res.* **2008**, *86*, 2973–2983. [PubMed]

16. Liu, L.; Li, C.J.; Lu, Y.; Zong, X.G.; Luo, C.; Sun, J.; Guo, L.J. Baclofen mediates neuroprotection on hippocampal CA1 pyramidal cells through the regulation of autophagy under chronic cerebral hypoperfusion. *Sci. Rep.* **2015**, *5*, 14474. [CrossRef] [PubMed]

17. Wei, X.W.; Yan, H.; Xu, B.; Wu, Y.P.; Li, C.; Zhang, G.Y. Neuroprotection of co-activation of GABA receptors by preventing caspase-3 denitrosylation in KA-induced seizures. *Brain. Res. Bull.* **2012**, *88*, 617–623. [CrossRef] [PubMed]

18. Parvez, M.K. Natural or plant products for the treatment of neurological disorders: Current knowledge. *Curr. Drug. Metab.* **2018**, *19*, 424–428. [CrossRef] [PubMed]

19. Okada, T.; Sugishita, T.; Murakami, T.; Murai, H.; Saikusa, T.; Horino, T.; Onoda, A.; Kajimoto, O.; Takahashi, R.; Takahashi, T. Effect of the defatted rice germ enriched with GABA for sleeplessness, depression, autonomic disorder by oral administration. *J. Jpn. Soc. Food Sci.* **2000**, *47*, 596–603. [CrossRef]

20. Chuang, C.Y.; Shi, Y.C.; You, H.P.; Lo, Y.H.; Pan, T.M. Antidepressant effect of GABA-rich monascus-fermented product on forced swimming rat model. *J. Agric. Food Chem.* **2011**, *59*, 3027–3034. [CrossRef]

21. Yamatsu, A.; Yamashita, Y.; Pandharipande, T.; Maru, I.; Kim, M. Effect of oral γ-aminobutyric acid (GABA) administration on sleep and its absorption in humans. *Food Sci. Biotechnol.* **2016**, *25*, 547–551. [CrossRef]

22. Kim, S.; Jo, K.; Hong, K.B.; Han, S.H.; Suh, H.J. GABA and l-theanine mixture decreases sleep latency and improves NREM sleep. *Pharm. Biol.* **2019**, *57*, 65–73. [CrossRef]

23. Abdou, A.M.; Higashiguchi, S.; Horie, K.; Kim, M.; Hatta, H.; Yokogoshi, H. Relaxation and immunity enhancement effect of gamma-aminobutyric acid (GABA) administration in humans. *BioFactors* **2006**, *26*, 201–208. [CrossRef]

24. Seo, Y.C.; Choi, W.Y.; Kim, J.S.; Lee, C.G.; Ahn, J.H.; Cho, H.Y.; Lee, S.H.; Cho, J.S.; Joo, S.J.; Lee, H.Y. Enhancement of the cognitive effects of GABA from monosodium glutamate fermentation by *Lactobacillus sakei* B2-16. *Food Biotechnol.* **2012**, *26*, 29–44. [CrossRef]

25. Reid, S.N.S.; Ryu, J.K.; Kim, Y.; Jeon, B.H. The effects of fermented *Laminaria japonica* on short-term working memory and physical fitness in the elderly. *Evid. Based Complement. Altern. Med.* **2018**, *2018*, 8109621.

26. Reid, S.N.S.; Ryu, J.K.; Kim, Y.S.; Jeon, B.H. GABA-enriched fermented *Laminaria japonica* improves cognitive impairment and neuroplasticity in scopolamine- and ethanol-induced dementia model mice. *Nutr. Res. Pract.* **2018**, *12*, 199–207. [CrossRef] [PubMed]

27. Choi, W.C.; Reid, S.N.S.; Ryu, J.K.; Kim, Y.; Jo, Y.H.; Jeon, B.H. Effects of γ-aminobutyric acid-enriched fermented sea tangle (*Laminaria japonica*) on brain derived neurotrophic factor-related muscle growth and lipolysis in middle aged women. *Algae* **2016**, *31*, 175–187. [CrossRef]

28. Schellack, N.; Naicker, P. Hypertension: A review of antihypertensive medication, past and present. *S. Afr. Pharm. J.* **2015**, *82*, 17–25.

29. Murray, B.A.; FitzGerald, R.J. Angiotensin converting enzyme inhibitory peptides derived from food proteins: Biochemistry, bioactivity and production. *Curr. Pharm. Des.* **2007**, *13*, 773–791. [CrossRef]

30. Nejati, F.; Rizzello, C.G.; Cagno, R.D.; Sheikh-Zeinoddin, M.; Diviccaro, A.; Minervini, F.; Gobbetti, M. Manufacture of a functional fermented milk enriched of angiotensin-I converting enzyme (ACE)-inhibitory peptides and γ-amino butyric acid (GABA). *Food Sci. Technol.* **2013**, *51*, 183–189. [CrossRef]

31. Tung, Y.T.; Lee, B.H.; Liu, C.F.; Pan, T.M. Optimization of culture condition for ACEI and GABA production by lactic acid bacteria. *J. Food Sci.* **2011**, *76*, M585–M591. [CrossRef]

32. Jang, E.K.; Kim, N.Y.; Ahn, H.J.; Ji, G.E. γ-Aminobutyric acid (GABA) production and angiotensin-I converting enzyme (ACE) inhibitory activity of fermented soybean containing sea tangle by the co-culture of *Lactobacillus brevis* with *Aspergillus oryzae*. *J. Microbiol. Biotechnol.* **2015**, *25*, 1315–1320. [CrossRef]

33. Sang, V.T.; Uyen, L.P.; Hung, N.D. The increased gamma-aminobutyric acid content by optimizing fermentation conditions of bacteria from kimchi and investigation of its biological activities. *EurAsian J. Biosci.* **2018**, *12*, 369–376.

34. Torino, M.I.; Limón, R.I.; Martínez-Villaluenga, C.; Mäkinen, S.; Pihlanto, A.; Vidal-Valverde, C.; Frias, J. Antioxidant and antihypertensive properties of liquid and solid state fermented lentils. *Food Chem.* **2013**, *136*, 1030–1037. [CrossRef]

35. Kimura, M.; Hayakawa, K.; Sansawa, H. Involvement of gamma-aminobutyric acid (GABA) B receptors in the hypotensive effect of systemically administered GABA in spontaneously hypertensive rats. *Jpn. J. Pharmacol.* **2002**, *89*, 388–394. [CrossRef] [PubMed]

36. Hayakawa, K.; Kimura, M.; Kasaha, K.; Matsumoto, K.; Sansawa, H.; Yamori, Y. Effect of a gamma-aminobutyric acid-enriched dairy product on the blood pressure of spontaneously hypertensive and normotensive Wistar-Kyoto rats. *Br. J. Nutr.* **2004**, *92*, 411–417. [CrossRef] [PubMed]

37. Matsubara, F.; Ueno, H.; Kentaro, T.; Tadano, K.; Suyama, T.; Imaizumi, K.; Suzuki, T.; Magata, K.; Kikuchi, N. Effects of GABA supplementation on blood pressure and safety in adults with mild hypertension. *Jpn. Pharmacol. Ther.* **2002**, *30*, 963–972.

38. Akama, K.; Kanetou, J.; Shimosaki, S.; Kawakami, K.; Tsuchikura, S.; Takaiwa, F. Seed-specific expression of truncated OsGAD2 produces GABA-enriched rice grains that influence a decrease in blood pressure in spontaneously hypertensive rats. *Transgenic Res.* **2009**, *18*, 865–876. [CrossRef]

39. Ebizuka, H.; Ihara, M.; Arita, M. Antihypertensive effect of pre-germinated brown rice in spontaneously hypertensive rats. *Food Sci. Technol. Res.* **2009**, *15*, 625–630. [CrossRef]

40. Kawakami, K.; Yamada, K.; Yamada, T.; Nabika, T.; Nomura, M. Antihypertensive effect of γ-aminobutyric acid-enriched brown rice on spontaneously hypertensive rats. *J. Nutr. Sci. Vitaminol. (Tokyo)* **2018**, *64*, 56–62. [CrossRef]

41. Nishimura, M.; Yoshida, S.; Haramoto, M.; Mizuno, H.; Fukuda, T.; Kagami-Katsuyama, H.; Tanaka, A.; Ohkawara, T.; Sato, Y.; Nishihira, J. Effects of white rice containing enriched gamma-aminobutyric acid on blood pressure. *J. Tradit. Complement. Med.* **2016**, *6*, 66–71. [CrossRef]

42. Tsai, C.C.; Chiu, T.H.; Ho, C.Y.; Lin, P.P.; Wu, T.Y. Effects of anti-hypertension and intestinal microflora of spontaneously hypertensive rats fed gammaaminobutyric acid-enriched Chingshey purple sweet potato fermented milk by lactic acid bacteria. *Afr. J. Microbiol. Res.* **2013**, *7*, 932–940.

43. Lin, P.P.; Hsieh, Y.M.; Kuo, W.W.; Lin, C.C.; Tsai, F.J.; Tsai, C.H.; Huang, C.Y.; Tsai, C.C. Inhibition of cardiac hypertrophy by probiotic-fermented purple sweet potato yogurt in spontaneously hypertensive rat hearts. *Int. J. Mol. Med.* **2012**, *30*, 1365–1375. [CrossRef]

44. Suwanmanon, K.; Hsieh, P.C. Effect of γ-aminobutyric acid and nattokinase-enriched fermented beans on the blood pressure of spontaneously hypertensive and normotensive Wistar-Kyoto rats. *J. Food Drug. Anal.* **2014**, *22*, 485–491. [CrossRef] [PubMed]

45. Aoki, H.; Furuya, Y.; Endo, Y.; Fujimoto, K. Effect of gamma-aminobutyric acid-enriched tempeh-like fermented soybean (GABA-Tempeh) on the blood pressure of spontaneously hypertensive rats. *Biosci. Biotechnol. Biochem.* **2003**, *67*, 1806–1808. [CrossRef] [PubMed]

46. Yoshimura, M.; Toyoshi, T.; Sano, A.; Izumi, T.; Fujii, T.; Konishi, C.; Inai, S.; Matsukura, C.; Fukuda, N.; Ezura, H.; et al. Antihypertensive effect of a gamma-aminobutyric acid rich tomato cultivar 'DG03-9' in spontaneously hypertensive rats. *J. Agric. Food Chem.* **2010**, *58*, 615–619. [CrossRef] [PubMed]

47. Becerra-Tomás, N.; Guasch-Ferré, M.; Quilez, J.; Merino, J.; Ferré, R.; Díaz-López, A.; Bulló, M.; Hernández-Alonso, P.; Palau-Galindo, A.; Salas-Salvadó, J. Effect of functional bread rich in potassium, γ-aminobutyric acid and angiotensin-converting enzyme inhibitors on blood pressure, glucose metabolism and endothelial function: A double-blind randomized crossover clinical trial. *Medicine (Baltimore)* **2015**, *94*, e1807. [CrossRef] [PubMed]

48. Kharroubi, A.T.; Darwish, H.M. Diabetes mellitus: The epidemic of the century. *World J. Diabetes* **2015**, *6*, 850–867. [CrossRef] [PubMed]

49. Li, G.; Zhang, P.; Wang, J.; Gregg, E.W.; Yang, W.; Gong, Q. The long-term effect of life style interventions to prevent diabetes in the China Da Qing diabetes prevention study: A 20-year follow-up study. *Lancet* **2008**, *371*, 1783–1789. [CrossRef]

50. Babiker, A.; Dubayee, M.A. Anti-diabetic medications: How to make a choice? *Sudan J. Paediatr.* **2017**, *17*, 11–20. [CrossRef] [PubMed]

51. Soltani, N.; Qiu, H.; Aleksic, M.; Glinka, Y.; Zhao, F.; Liu, R.; Li, Y.; Zhang, N.; Chakrabarti, R.; Ng, T.; et al. GABA exerts protective and regenerative effects on islet beta cells and reverses diabetes. *Proc. Natl. Acad. Sci. USA* **2011**, *108*, 11692–11697. [CrossRef] [PubMed]

52. Untereiner, A.; Abdo, S.; Bhattacharjee, A.; Gohil, H.; Pourasgari, F.; Ibeh, N.; Lai, M.; Batchuluun, B.; Wong, A.; Khuu, N.; et al. GABA promotes β-cell proliferation, but does not overcome impaired glucose homeostasis associated with diet-induced obesity. *FASEB J.* **2019**, *33*, 3968–3984. [CrossRef]

53. Liu, W.; Son, D.O.; Lau, H.K.; Zhou, Y.; Prud'homme, G.J.; Jin, T.; Wang, Q. Combined oral administration of GABA and DPP-4 inhibitor prevents beta cell damage and promotes beta cell regeneration in mice. *Front. Pharmacol.* **2017**, *8*, 362. [CrossRef]

54. Bansal, P.; Wang, S.; Liu, S.; Xiang, Y.Y.; Lu, W.Y.; Wang, Q. GABA coordinates with insulin in regulating secretory function in pancreatic INS-1 β-cells. *PLoS ONE* **2011**, *6*, e26225. [CrossRef] [PubMed]

55. Tian, J.; Dang, H.N.; Yong, J.; Chui, W.S.; Dizon, M.P.; Yaw, C.K.; Kaufman, D.L. Oral treatment with γ-aminobutyric acid improves glucose tolerance and insulin sensitivity by inhibiting inflammation in high fat diet-fed mice. *PLoS ONE* **2011**, *6*, e25338. [CrossRef] [PubMed]

56. Huang, C.Y.; Kuo, W.W.; Wang, H.F.; Lin, C.J.; Lin, Y.M.; Chen, J.L.; Kuo, C.H.; Chen, P.K.; Lin, J.Y. GABA tea ameliorates cerebral cortex apoptosis and autophagy in streptozotocin-induced diabetic rats. *J. Funct. Foods* **2014**, *6*, 534–544. [CrossRef]

57. Imam, M.U.; Azmi, N.H.; Bhanger, M.I.; Ismail, N.; Ismail, M. Antidiabetic properties of germinated brown rice: A systematic review. *J. Evid. Based Complement. Altern. Med.* **2012**, *2012*, 816501.

58. Sivamaruthi, B.S.; Kesika, P.; Prasanth, M.I.; Chaiyasut, C. A mini review on antidiabetic properties of fermented foods. *Nutrients* **2018**, *10*, 1973. [CrossRef] [PubMed]

59. Hagiwara, H.; Seki, T.; Ariga, T. The effect of pre-germinated brown rice intake on blood glucose and PAI-1 levels in streptozotocin-induced diabetic rats. *Biosci. Biotechnol. Biochem.* **2004**, *68*, 444–447. [CrossRef] [PubMed]

60. Torimitsu, M.; Nagase, R.; Yanagi, M.; Homma, M.; Sasai, Y.; Ito, Y.; Hayamizu, K.; Nonaka, S.; Hosono, T.; Kise, M.; et al. Replacing white rice with pre-germinated brown rice mildly ameliorates hyperglycemia and imbalance of adipocytokine levels in type 2 diabetes model rats. *J. Nutr. Sci. Vitaminol. (Tokyo)* **2010**, *56*, 287–292. [CrossRef]

61. Adamu, H.A.; Imam, M.U.; Ooi, D.J.; Esa, N.M.; Rosli, R.; Ismail, M. Perinatal exposure to germinated brown rice and its gamma amino-butyric acid-rich extract prevents high fat diet-induced insulin resistance in first generation rat offspring. *Food Nutr. Res.* **2016**, *60*, 30209. [CrossRef]

62. Chung, S.I.; Jin, X.; Kang, M.Y. Enhancement of glucose and bone metabolism in ovariectomized rats fed with germinated pigmented rice with giant embryo (*Oryza sativa* L. cv. Keunnunjami). *Food Nutr. Res.* **2019**, *63*. [CrossRef]

63. Shang, W.; Si, X.; Zhou, Z.; Strappe, P.; Blanchard, C. Wheat bran with enriched gamma-aminobutyric acid attenuates glucose intolerance and hyperinsulinemia induced by a high-fat diet. *Food Funct.* **2018**, *9*, 2820–2828. [CrossRef]

64. Si, X.; Shang, W.; Zhou, Z.; Shui, G.; Lam, S.M.; Blanchard, C.; Strappe, P. Gamma-aminobutyric acid enriched rice bran diet attenuates insulin resistance and balances energy expenditure via modification of gut microbiota and short-chain fatty acids. *J. Agric. Food Chem.* **2018**, *66*, 881–890. [CrossRef] [PubMed]

65. Ito, Y.; Mizukuchi, A.; Kise, M.; Aoto, H.; Yamamoto, S.; Yoshihara, R.; Yokoyama, J. Postprandial blood glucose and insulin responses to pre-germinated brown rice in healthy subjects. *J. Med. Investig.* **2005**, *52*, 159–164. [CrossRef]

66. Hsu, T.F.; Kise, M.; Wang, M.F.; Ito, Y.; Yang, M.D.; Aoto, H.; Yoshihara, R.; Yokoyama, J.; Kunii, D.; Yamamoto, S. Effects of pre-germinated brown rice on blood glucose and lipid levels in free-living patients with impaired fasting glucose or type 2 diabetes. *J. Nutr. Sci. Vitaminol. (Tokyo)* **2008**, *54*, 163–168. [CrossRef] [PubMed]

67. Hayakawa, T.; Suzuki, S.; Kobayashi, S.; Fukutomi, T.; Ide, M.; Ohno, T.; Ohkouchi, M.; Taki, M.; Miyamoto, T.; Nimura, T. Effect of pre-germinated brown rice on metabolism of glucose and lipid in patients with diabetes mellitus type 2. *J. Jpn. Assoc. Rural Med.* **2009**, *58*, 438–446. [CrossRef]

68. Tamaya, K.; Matsui, T.; Toshima, A.; Noguchi, M.; Ju, Q.; Miyata, Y.; Tanaka, T.; Tanaka, K. Suppression of blood glucose level by a new fermented tea obtained by tea-rolling processing of loquat (*Eriobotrya japonica*) and green tea leaves in disaccharide-loaded Sprague-Dawley rats. *J. Sci. Food Agric.* **2010**, *90*, 779–783. [CrossRef] [PubMed]

69. Lee, S.Y.; Park, S.L.; Nam, Y.D.; Lee, S.H. Anti-diabetic effects of fermented green tea in KK-Ay diabetic mice. *Korean J. Food Sci. Technol.* **2013**, *45*, 488–494. [CrossRef]

70. Yeap, S.K.; Ali, N.M.; Yusof, H.M.; Alitheen, N.B.; Beh, B.K.; Ho, W.Y.; Koh, S.P.; Long, K. Antihyperglycemic effects of fermented and nonfermented mung bean extracts on alloxan-induced-diabetic mice. *J. Biomed. Biotechnol.* **2012**, *2012*, 285430. [CrossRef]

71. Chen, L.; Zhao, H.; Zhang, C.; Lu, Y.; Zhu, X.; Lu, Z. γ-Aminobutyric acid-rich yogurt fermented by *Streptococcus salivarius* subsp. thermophiles fmb5 apprars to have anti-diabetic effect on streptozotocin-induced diabetic mice. *J. Funct. Foods* **2016**, *20*, 267–275. [CrossRef]

72. Nam, H.; Jung, H.; Karuppasamy, S.; Park, Y.S.; Cho, Y.S.; Lee, J.Y.; Seong, S.I.; Suh, J.G. Anti-diabetic effect of the soybean extract fermented by *Bacillus subtilis* MORI in db/db mice. *Food Sci. Biotechnol.* **2012**, *21*, 1669–1676. [CrossRef]

73. Song, K.; Song, I.B.; Gu, H.J.; Na, J.; Kim, S.; Yang, H.S.; Lee, S.C.; Huh, C.; Kwon, J. Anti-diabetic effect of fermented milk containing conjugated linoleic acid on type II diabetes mellitus. *Korean J. Food Sci. Anim. Resour.* **2016**, *36*, 170–177. [CrossRef]

74. Ostadrahimi, A.; Taghizadeh, A.; Mobasseri, M.; Farrin, N.; Payahoo, L.; Gheshlaghi, Z.B.; Vahedjabbari, M. Effect of probiotic fermented milk (kefir) on glycemic control and lipid profile in type 2 diabetic patients: A randomized double-blind placebo-controlled clinical trial. *Iran. J. Public Health* **2015**, *44*, 228–237.

75. Alihosseini, N.; Moahboob, S.A.; Farrin, N.; Mobasseri, M.; Taghizadeh, A.; Ostadrahimi, A.R. Effect of probiotic fermented milk (kefir) on serum level of insulin and homocysteine in type 2 diabetes patients. *Acta Endocrinol.* **2017**, *13*, 431–436. [CrossRef] [PubMed]

76. Ribas, V.; García-Ruiz, C.; Fernández-Checa, J.C. Mitochondria, cholesterol and cancer cell metabolism. *Clin. Transl. Med.* **2016**, *5*, 22. [CrossRef] [PubMed]

77. Oh, C.H.; Oh, S.H. Effects of germinated brown rice extracts with enhanced levels of GABA on cancer cell proliferation and apoptosis. *J. Med. Food* **2004**, *7*, 19–23. [CrossRef] [PubMed]

78. Schuller, H.M.; Al-Wadei, H.A.; Majidi, M. Gamma-aminobutyric acid, a potential tumor suppressor for small airway-derived lung adenocarcinoma. *Carcinogenesis* **2008**, *29*, 1979–1985. [CrossRef]

79. Huang, Q.; Liu, C.; Wang, C.; Hu, Y.; Qiu, L.; Xu, P. Neurotransmitter γ-aminobutyric acid-mediated inhibition of the invasive ability of cholangiocarcinoma cells. *Oncol. Lett.* **2011**, *2*, 519–523. [CrossRef]

80. Song, L.; Du, A.; Xiong, Y.; Jiang, J.; Zhang, Y.; Tian, Z.; Yan, H. γ-Aminobutyric acid inhibits the proliferation and increases oxaliplatin sensitivity in human colon cancer cells. *Tumour. Biol.* **2016**, *37*, 14885–14894. [CrossRef]

81. Al-Wadei, H.A.; Al-Wadei, M.H.; Ullah, M.F.; Schuller, H.M. Celecoxib and GABA cooperatively prevent the progression of pancreatic cancer in vitro and in xenograft models of stress-free and stress-exposed mice. *PLoS ONE* **2012**, *7*, e43376. [CrossRef]

82. Tatsuta, M.; Iishi, H.; Baba, M.; Nakaizumi, A.; Ichii, M.; Taniguchi, H. Inhibition by gamma-amino-n-butyric acid and baclofen of gastric carcinogenesis induced by N-methyl-N'-nitro-N-nitrosoguanidine in Wistar rats. *Cancer Res.* **1990**, *50*, 4931–4934.

83. Chen, Z.A.; Bao, M.Y.; Xu, Y.F.; Zha, R.P.; Shi, H.B.; Chen, T.Y.; He, X.H. Suppression of human liver cancer cell migration and invasion via the GABAA receptor. *Cancer Biol. Med.* **2012**, *9*, 90–98. [PubMed]

84. Zhang, M.; Gong, Y.; Assy, N.; Minuk, Y. Increased GABAergic activity inhibits a-fetoprotein mRNA expression and the proliferation activity of the HepG2 human hepatocellular carcinoma cell line. *J. Hepatol.* **2000**, *32*, 85–91. [CrossRef]

85. Phaniendra, A.; Jestadi, D.B.; Periyasamy, L. Free radicals: Properties, sources, targets, and their implication in various diseases. *Indian J. Clin. Biochem.* **2015**, *30*, 11–26. [CrossRef] [PubMed]

86. Lobo, V.; Patil, A.; Phatak, A.; Chandra, N. Free radicals, antioxidants and functional foods: Impact on human health. *Pharmacogn. Rev.* **2010**, *4*, 118–126. [CrossRef] [PubMed]

87. Deng, Y.; Xu, L.; Zeng, X.; Li, Z.; Qin, B.; He, N. New perspective of GABA as an inhibitor of formation of advanced lipoxidation end-products: it's interaction with malondiadehyde. *J. Biomed. Nanotechnol.* **2010**, *6*, 318–324. [CrossRef] [PubMed]

88. Deng, Y.; Wang, W.; Yu, P.; Xi, Z.; Xu, L.; Li, X.; He, N. Comparison of taurine, GABA, Glu, and Asp as scavengers of malondialdehyde in vitro and in vivo. *Nanoscale Res. Lett.* **2013**, *8*, 190. [CrossRef] [PubMed]

89. Tang, X.; Yu, R.; Zhou, Q.; Jiang, S.; Le, G. Protective effects of γ-aminobutyric acid against H_2O_2-induced oxidative stress in RIN-m5F pancreatic cells. *Nutr. Metab. (Lond.)* **2018**, *15*, 60. [CrossRef]

90. Zhu, Z.; Shi, Z.; Xie, C.; Gong, W.; Hu, Z.; Peng, Y. A novel mechanism of gamma-aminobutyric acid (GABA) protecting human umbilical vein endothelial cells (HUVECs) against H_2O_2-induced oxidative injury. *Comp. Biochem. Physiol. C Toxicol. Pharmacol.* **2019**, *217*, 68–75. [CrossRef]

91. El-Hady, A.M.A.; Gewefel, H.S.; Badawi, M.A.; Eltahawy, N.A. Gamma-aminobutyric acid ameliorates gamma rays-induced oxidative stress in the small intestine of rats. *J. Basic Appl. Zool.* **2017**, *78*, 2. [CrossRef]

92. Eltahawy, N.A.; Saada, H.N.; Hammad, A.S. Gamma amino butyric acid attenuates brain oxidative damage associated with insulin alteration in streptozotocin-treated rats. *Indian J. Clin. Biochem.* **2017**, *32*, 207–213. [CrossRef]

93. Lee, B.J.; Kim, J.S.; Kang, Y.M.; Lim, J.H.; Kim, Y.M.; Lee, M.S.; Jeong, M.H.; Ahn, C.B.; Je, J.Y. Antioxidant activity and γ-aminobutyric acid (GABA) content in sea tangle fermented by *Lactobacillus brevis* BJ20 isolated from traditional fermented foods. *Food Chem.* **2010**, *122*, 271–276. [CrossRef]

94. Md Zamri, N.D.; Imam, M.U.; Abd Ghafar, S.A.; Ismail, M. Antioxidative effects of germinated brown rice-derived extracts on H_2O_2-induced oxidative stress in HepG2 cells. *Evid. Based Complement. Altern. Med.* **2014**, *2014*, 371907. [CrossRef] [PubMed]

95. Phuapaiboon, P. Gamma-aminobutyric acid, total anthocyanin content and antioxidant activity of vinegar brewed from germinated pigmented rice. *Pakistan J. Nutr.* **2017**, *16*, 109–118.

96. Chen, L.; Deng, H.; Cui, H.; Fang, J.; Zuo, Z.; Deng, J.; Li, Y.; Wang, X.; Zhao, L. Inflammatory responses and inflammation-associated diseases in organs. *Oncotarget* **2017**, *9*, 7204–7218. [CrossRef] [PubMed]

97. Abdulkhaleq, L.A.; Assi, M.A.; Abdullah, R.; Zamri-Saad, M.; Taufiq-Yap, Y.H.; Hezmee, M. The crucial roles of inflammatory mediators in inflammation: A review. *Vet. World* **2018**, *11*, 627–635. [CrossRef] [PubMed]

98. Han, D.; Kim, H.Y.; Lee, H.J.; Shim, I.; Hahm, D.H. Wound healing activity of gamma-aminobutyric acid (GABA) in rats. *J. Microbiol. Biotechnol.* **2007**, *17*, 1661–1669. [PubMed]

99. Prud'homme, G.; Glinka, Y.; Wang, Q. GABA exerts anti-inflammatory and immunosuppressive effects (P5175). *J. Immunol.* **2013**, *190*, 68.

100. Choi, J.I.; Yun, I.H.; Jung, Y.J.; Lee, E.H.; Nam, T.J.; Kim, Y.M. Effects of γ-aminobutyric acid (gaba)-enriched sea tangle *Laminaria japonica* extract on lipopolysaccharide-induced inflammation in mouse macrophage (RAW 264.7) cells. *Fish Aquat. Sci.* **2012**, *15*, 293–297. [CrossRef]

101. Tuntipopipat, S.; Muangnoi, C.; Thiyajai, P.; Srichamnong, W.; Charoenkiatkul, S.; Praengam, K. A bioaccessible fraction of parboiled germinated brown rice exhibits a higher anti-inflammatory activity than that of brown rice. *Food Funct.* **2015**, *6*, 1480–1488. [CrossRef]

102. Scandolera, A.; Hubert, J.; Humeau, A.; Lambert, C.; De Bizemont, A.; Winkel, C.; Kaouas, A.; Renault, J.H.; Nuzillard, J.M.; Reynaud, R. GABA and GABA-alanine from the red microalgae *Rhodosorus marinus* exhibit a significant neuro-soothing activity through inhibition of neuro-inflammation mediators and positive regulation of TRPV1-related skin sensitization. *Mar. Drugs* **2018**, *16*, 96. [CrossRef]

103. Sokovic Bajic, S.; Djokic, J.; Dinic, M.; Veljovic, K.; Golic, N.; Mihajlovic, S.; Tolinacki, M. GABA-producing natural dairy isolate from artisanal zlatar cheese attenuates gut inflammation and strengthens gut epithelial barrier in vitro. *Front. Microbiol.* **2019**, *10*, 527. [CrossRef]

104. Mau, J.L.; Chiou, S.Y.; Hsu, C.A.; Tsai, H.L.; Lin, S.D. Antimutagenic and antimicrobial activities of γ-aminobutyric acid (Gaba) tea extract. *Int. Conf. Nutr. Food Sci.* **2012**, *39*, 178–182.

105. Dagorn, A.; Hillion, M.; Chapalain, A.; Lesouhaitier, O.; Duclairoir Poc, C.; Vieillard, J.; Chevalier, S.; Taupin, L.; Le Derf, F.; Feuilloley, M.G. Gamma-aminobutyric acid acts as a specific virulence regulator in *Pseudomonas aeruginosa*. *Microbiology* **2013**, *159*, 339–351. [CrossRef] [PubMed]

106. Kim, J.K.; Kim, Y.S.; Lee, H.M.; Jin, H.S.; Neupane, C.; Kim, S.; Lee, S.H.; Min, J.J.; Sasai, M.; Jeong, J.H.; et al. GABAergic signaling linked to autophagy enhances host protection against intracellular bacterial infections. *Nat. Commun.* **2018**, *9*, 4184. [CrossRef] [PubMed]

107. Galli, S.J.; Tsai, M. IgE and mast cells in allergic disease. *Nat. Med.* **2012**, *18*, 693–704. [CrossRef] [PubMed]

108. Hori, A.; Hara, T.; Honma, K.; Joh, T. Suppressive effect of γ-aminobutyric acid (GABA) on histamine release in rat basophilic RBL-2H3 cells. *Bull. Fac. Agric. Niigata Univ.* **2008**, *61*, 47–51.

109. Kawasaki, A.; Hara, T.; Joh, T. Inhibitory effect of γ-aminobutyric acid (GABA) on histamine release from rat basophilic leukemia RBL-2H3 cells and rat peritoneal exudate cells. *Nippon Shokuhin Kagaku Kogaku Kaishi* **2014**, *61*, 362–366. [CrossRef]

110. Hokazono, H.; Omori, T.; Ono, K. Effects of single and combined administration of fermented barley extract and γ -aminobutyric acid on the development of atopic dermatitis in NC/Nga mice. *Biosci. Biotechnol. Biochem.* **2010**, *74*, 135–139. [CrossRef] [PubMed]

111. Nagy, L.E. Molecular aspects of alcohol metabolism: Transcription factors involved in early ethanol-induced liver injury. *Annu. Rev. Nutr.* **2004**, *24*, 55–78. [CrossRef]

112. Oh, S.H.; Soh, J.R.; Cha, Y.S. Germinated brown rice extract shows a nutraceutical effect in the recovery of chronic alcohol-related symptoms. *J. Med. Food* **2003**, *6*, 115–121. [CrossRef]

113. Lee, B.J.; Senevirathne, M.; Kim, J.S.; Kim, Y.M.; Lee, M.S.; Jeong, M.H.; Kang, Y.M.; Kim, J.I.; Nam, B.H.; Ahn, C.B.; et al. Protective effect of fermented sea tangle against ethanol and carbon tetrachloride-induced hepatic damage in Sprague-Dawley rats. *Food Chem. Toxicol.* **2010**, *48*, 1123–1128. [CrossRef]

114. Cha, J.Y.; Lee, B.J.; Je, J.Y.; Kang, Y.M.; Kim, Y.M.; Cho, Y.S. GABA-enriched fermented *Laminaria japonica* protects against alcoholic hepatotoxicity in Sprague-Dawley rats. *Fish Aquat. Sci.* **2011**, *14*, 79–88. [CrossRef]

115. Kang, Y.M.; Qian, Z.J.; Lee, B.J.; Kim, Y.M. Protective effect of GABA-enriched fermented sea tangle against ethanol-induced cytotoxicity in HepG2 Cells. *Biotechnol. Bioprocess E* **2011**, *16*, 966. [CrossRef]

116. Makris, K.; Spanou, L. Acute kidney injury: Definition, pathophysiology and clinical phenotypes. *Clin. Biochem. Rev.* **2016**, *37*, 85–98. [PubMed]

117. Kim, H.Y.; Yokozawa, T.; Nakagawa, T.; Sasaki, S. Protective effect of gamma-aminobutyric acid against glycerol-induced acute renal failure in rats. *Food Chem. Toxicol.* **2004**, *42*, 2009–2014. [CrossRef] [PubMed]

118. Sasaki, S.; Yokozawa, T.; Cho, E.J.; Oowada, S.; Kim, M. Protective role of gamma-aminobutyric acid against chronic renal failure in rats. *J. Pharm. Pharmacol.* **2006**, *58*, 1515–1525. [CrossRef]

119. Sasaki, S.; Tohda, C.; Kim, M.; Yokozawa, T. Gamma-aminobutyric acid specifically inhibits progression of tubular fibrosis and atrophy in nephrectomized rats. *Biol. Pharm. Bull.* **2007**, *30*, 687–691. [CrossRef] [PubMed]

120. Ali, B.H.; Al-Salam, S.; Al Za'abi, M.; Al Balushi, K.A.; AlMahruqi, A.S.; Beegam, S.; Al-Lawatia, I.; Waly, M.I.; Nemmar, A. Renoprotective effects of gamma-aminobutyric acid on cisplatin-induced acute renal injury in rats. *Basic Clin. Pharmacol. Toxicol.* **2015**, *116*, 62–68. [CrossRef] [PubMed]

121. Talebi, N.; Nematbakhsh, M.; Monajemi, R.; Mazaheri, S.; Talebi, A.; Vafapour, M. The protective effect of γ-aminobutyric acid on kidney injury induced by renal ischemia-reperfusion in ovariectomized estradiol-treated rats. *Int. J. Prev. Med.* **2016**, *7*, 6.

122. Saada, H.N.; Eltahawy, N.A.; Hammad, A.S.; Morcos, N.Y.S. Gamma amino butyric acid attenuates liver and kidney damage associated with insulin alteration in γ-irradiated and streptozotocin-treated rats. *Arab. J. Nucl. Sci. Appl.* **2016**, *49*, 138–150.

123. Vafapour, M.; Nematbakhsh, M.; Monajemi, R.; Mazaheri, S.; Talebi, A.; Talebi, N.; Shirdavani, S. Effect of γ-aminobutyric acid on kidney injury induced by renal ischemia-reperfusion in male and female rats: Gender-related difference. *Adv. Biomed. Res.* **2015**, *4*, 158.

124. Chen, Z.; Xie, J.; Wang, B.; Tang, J. Effect of γ-aminobutyric acid on digestive enzymes, absorption function, and immune function of intestinal mucosa in heat-stressed chicken. *Poult. Sci.* **2014**, *93*, 2490–2500. [CrossRef] [PubMed]

125. Chen, Z.; Xie, J.; Hu, M.Y.; Tang, J.; Shao, Z.F.; Li, M.H. Protective effects of γ-aminobutyric acid (gaba) on the small intestinal mucosa in heat-stressed wenchang chicken. *J. Anim. Plant Sci.* **2015**, *25*, 78–87.

126. Chen, S.; Tan, B.; Xia, Y.; Liao, S.; Wang, M.; Yin, J.; Wang, J.; Xiao, H.; Qi, M.; Bin, P.; et al. Effects of dietary gamma-aminobutyric acid supplementation on the intestinal functions in weaning piglets. *Food Funct.* **2019**, *10*, 366–378. [CrossRef] [PubMed]

127. Xie, M.; Chen, H.H.; Nie, S.P.; Yin, J.Y.; Xie, M.Y. Gamma-aminobutyric acid increases the production of short-chain fatty acids and decreases pH values in mouse colon. *Molecules* **2017**, *22*, 653.

128. Kubota, A.; Kobayashi, M.; Sarashina, S.; Takeno, R.; Yasuda, G.; Narumi, K.; Furugen, A.; Takahashi-Suzuki, N.; Iseki, K. Gamma-aminobutyric acid (GABA) attenuates ischemia reperfusion-induced alterations in intestinal immunity. *Biol. Pharm. Bull.* **2018**, *41*, 1874–1878. [CrossRef] [PubMed]

129. Jiang, T.; Yue, Y.; Li, F. Improvement of gamma-aminobutyric acid on intestinal mucosalbarrier injury of colitis induced by 2,4,6-trinitrobenzene sulfonic acid and alcohol. *Herald Med.* **2018**, *37*, 931–938.

130. Yang, Y.; Lian, Y.T.; Huang, S.Y.; Yang, Y.; Cheng, L.X.; Liu, K. GABA and topiramate inhibit the formation of human macrophage-derived foam cells by modulating cholesterol-metabolism-associated molecules. *Cell. Physiol. Biochem.* **2014**, *33*, 1117–1129. [CrossRef] [PubMed]

131. Yang, H.; Xing, R.; Liu, S.; Yu, H.; Li, P. Analysis of the protective effects of γ-aminobutyric acid during fluoride-induced hypothyroidism in male Kunming mice. *Pharm. Biol.* **2019**, *57*, 29–37. [CrossRef] [PubMed]

132. Powers, M.E.; Borst, S.E.; McCoy, S.C.; Conway, R.; Yarrow, J.F. The effects of gamma aminobutyric acid on growth hormone secretion at rest and following exercise. *Med. Sci. Sports Exerc.* **2003**, *35*, S271. [CrossRef]

133. Tujioka, K.; Okuyama, S.; Yokogoshi, H.; Fukaya, Y.; Hayase, K.; Horie, K.; Kim, M. Dietary gamma-aminobutyric acid affects the brain protein synthesis rate in young rats. *Amino Acids* **2007**, *32*, 255–260. [CrossRef] [PubMed]

134. Tujioka, K.; Ohsumi, M.; Horie, K.; Kim, M.; Hayase, K.; Yokogoshi, H. Dietary gamma-aminobutyric acid affects the brain protein synthesis rate in ovariectomized female rats. *J. Nutr. Sci. Vitaminol. (Tokyo)* **2009**, *55*, 75–80. [CrossRef] [PubMed]

135. Sakashita, M.; Nakamura, U.; Horie, N.; Yokoyama, Y.; Kim, M.; Fujita, S. Oral supplementation using gamma-aminobutyric acid and whey protein improves whole body fat-free mass in men after resistance training. *J. Clin. Med. Res.* **2019**, *11*, 428–434. [CrossRef] [PubMed]

136. Droge, W. Free radicals in the physiological control of cell function. *Physiol. Rev.* **2002**, *82*, 47–95. [CrossRef] [PubMed]

137. Pacher, P.; Beckman, J.S.; Liaudet, L. Nitric oxide and peroxynitrite in health and disease. *Physiol. Rev.* **2007**, *87*, 315–424. [CrossRef] [PubMed]

138. Pham-Huy, L.A.; He, H.; Pham-Huy, C. Free radicals, antioxidants in disease and health. *Int. J. Biomed. Sci.* **2008**, *4*, 89–96. [PubMed]

139. Willcox, J.K.; Ash, S.L.; Catignani, G.L. Antioxidants and prevention of chronic disease. *Crit. Rev. Food. Sci. Nutr.* **2004**, *44*, 275–295. [CrossRef] [PubMed]

MDPI
St. Alban-Anlage 66
4052 Basel
Switzerland
Tel. +41 61 683 77 34
Fax +41 61 302 89 18
www.mdpi.com

Molecules Editorial Office
E-mail: molecules@mdpi.com
www.mdpi.com/journal/molecules